micro:bit in Wonderland

CODING & CRAFT

with the BBC micro:bit

Inspired by *Alice's Adventures in Wonderland*

For children aged 9+ and beginners of all ages

Foreword by The Micro:bit Educational Foundation

Learn to Code
#techalice

Dr Tracy Gardner & Elbrie de Kock

TECH AGE™**KIDS**

Published by Tech Age Kids. techagekids.com | hello@techagekids.com

Text, diagrams and photography ©Tech Age Kids Limited 2018.

Written by Dr Tracy Gardner & Elbrie de Kock.

Typeset and designed by Tech Age Kids.

ISBN 978-1-9997879-0-5

TRADEMARKS

BBC micro:bit and micro:bit are trademarks of © 2015 British Broadcasting Corporation. microbit.org

Tech Age™ is a trademark of Tech Age Kids Limited. techagekids.com

Microsoft and Microsoft Edge are trademarks of Microsoft Corporation. microsoft.com

LEGO® is a trademark of © 2018 The LEGO Group. lego.com

Google Chrome™ is a trademark of Google Inc. google.com

All trademarks are the property of their respective owners.

ACKNOWLEDGEMENTS

Screenshots made available by Microsoft MakeCode, make based on the open source project Microsoft Programming F

Illustrations by Sir John Tenniel from 'Alice's Adventures in

Book inspired by 'Alice's Adventures in Wonderland' by Lev

All photography produced by Tech Age Kids and may not be authors.

LIMIT OF LIABILITY

THE AUTHORS AND PUBLISHER DO NOT ACCEPT RESPONSIBILITY FOR ANY LOSS OR DAMAGE, ACCIDENT OR INJURY RESULTING DIRECTLY OR INDIRECTLY FROM USING THE INFORMATION CONTAINED IN THIS BOOK.

INTERNET ADDRESSES

All the internet addresses (URLs) and information given in this book were valid at the date of print. Please contact Tech Age Kids if you find any missing or incorrect information.

ABOUT THE AUTHORS

DR TRACY GARDNER

Tracy Gardner has a Computer Science PhD. She has worked as a software engineer and software architect, including working for 10 years at IBM. Tracy has two children and now focuses on introducing technology to the next generation. Tracy is a director of Tech Age Kids. She also develops educational content for the Raspberry Pi Foundation. Between 2014 to 2017, Tracy taught Computing to Key Stage 2 children (aged 7-11). She volunteers at a Code Club and Coder Dojo.

ELBRIE DE KOCK

Elbrie de Kock has an Interior Design degree and has worked in a number of different industries as a designer and digital marketer. Elbrie has three children. Her eldest son's passion for computer programming inspired her to find opportunities for kids to learn to code. She uses her creative background and newly developed technology skills to create projects that combine craft, coding and electronics. Elbrie is a director of Tech Age Kids and organises creative technology events for children and families. She volunteers at a Code Club and founded the local Coder Dojo.

ABOUT TECH AGE KIDS

Tech Age Kids is an online company that helps parents and educators find constructive and creative uses of technology for children and teens.

The company creates educational material and online content for techagekids.com, including approachable project ideas, news and reviews of the latest educational and creative technology products, as well as advice on digital parenting issues.

Tech Age Kids believes that modern children should develop skills in coding, electronics and design so that they can understand the present and shape the future. The company supports the STEAM (Science, Technology, Engineering, Art and Mathematics), Maker and Digital Making movements.

♣ Sign up to the weekly email for the latest articles from techagekids.com.

♣ Join the digital parenting group at facebook.com/techageparents to share experiences and learn from other parents.

♣ Follow on social media at Facebook, Twitter, Instagram and occasionally YouTube.

♣ Get in touch by email at hello@techagekids.com.

FOREWORD

The original BBC micro:bit project was part of the BBC's 2015 **Make It Digital** season. The micro:bit was first conceived by the BBC as being for children to inspire them toward coding and digital creativity. The original project was a genuine partnership led by the BBC but involving over 30 organisations, from small, regional charities through to major multinational corporations, all of whom shared the vision of encouraging children to become creators with digital tools rather than just users of technology. There were already tools on the market but many of these were expensive (making them inaccessible to all); hard to be able to use quickly and easily; or, not particularly versatile. The micro:bit was designed to be all of those things and, for the most part, does it very well.

One of the great things about having the BBC lead on the project is that they can tap into their massive collection of media assets to promote the micro:bit, so we had a plethora of materials created around programmes such as Doctor Who, Strictly Come Dancing etc. that helped to make the device appealing to the target audience of 11-12 year old children; however, the BBC project is now over; the reach of the micro:bit in terms of age of the students is now much broader; and, the global popularity of the micro:bit takes it into territories where people are not familiar with Strictly…and Doctor Who. Also, many of the materials were brilliant projects but lacked a coherent process of learning behind them.

The **Micro:bit Educational Foundation** is a not-for- profit that was set up with the support of some of the original BBC project partners. We are only funded through a small royalty received from the sale of each device so, when it comes to driving our mission of "making every child an inventor", we rely very much on the support and enthusiasm of our user community to help with promotion, accessories, training and resources. When we were presented with the draft of this book it was a delight to explore. It is coherent, concise and comprehensive. It takes the readers on a learning journey but in a way that is full of fun activities, set in the context of a timeless story that is known and loved around the world.

Coding is a hugely important skill, one that will be increasingly in demand as we go further into this century. Young people are facing a future that, if they have the right abilities, will be exciting and full of opportunity as a high proportion of the jobs that will exist in 20, 30, 40 years' time do not exist today and they are likely to change jobs and careers multiple times during their working lives. But it's not just coding that will make a different, so will encouraging creativity as we will need people who will come up with the products and services that will lead to employment and profits. The micro:bit does both – providing people with the ability to learn code from simple drag-and- drop blocks through to full text coding but also being able to fiddle, have a go and to be able to make mistakes without the fear of breaking it.

By the time the reader has worked their way through all the activities in this book they will have got a good grasp of the principles of coding and computational thinking but they will have made things, been inspired to have a go at coming up with their own ideas and, most importantly, had fun.

The Micro:bit Educational Foundation

ACKNOWLEDGEMENTS

We would like to thank everyone who tried out the projects at schools and events. A special thanks to all the families (including Bethany and the January family) who worked through the projects at home. Your feedback has been valuable in making the projects accessible and your enjoyment and light-bulb moments have shown us that developing coding and craft projects is a worthwhile endeavour.

We particularly want to thank Anne Wan for her invaluable editing skills and non-technical perspective on the book and Chris January, Sian January and Sean McManus for their superb technical reviews and insights.

And thanks to Ruby for taking us on regular walks where we had the chance to work through tricky problems and come up with creative ideas.

TRACY GARDNER

I really appreciate my family who have been very involved in shaping *micro:bit in Wonderland*. My children, Caleb and Reuben, discussed project ideas, tried out the book, really got into the spirit of the projects, made them theirs and suggested improvements (yes you were right!) Beanz provided a capable sounding board when trying to find the best way to explain concepts and make sure everything was technically accurate (any remaining errors are of course ours.)

Thanks also to Amanda Williams and Dr Julie Greer who asked me to teach Computing to primary school children and put me on the path that led to the original *micro:bit in Wonderland* projects.

And a huge thanks to Elbrie for turning my scrappy prototypes into awesome projects and helping me to make the projects accessible to beginners.

ELBRIE DE KOCK

A big thanks to my family for releasing me to learn new skills and work many hours to bring this book to life. Jake and Daniel have been keen testers of the projects and willing models for photographs. My husband, Marcus, for being an enthusiastic supporter and being patient when I turned our conservatory into a photo studio. My dad, le Roux, for teaching me to never be afraid to pursue something new.

A special thanks to Josh for nudging me on a new trajectory which led me to explore opportunities in the technology industry.

Many thanks to Tracy for patiently teaching me coding and technology concepts and working well together using our complimentary skills to bring the coding and craft projects to life.

'For, you see, so many out-of-the-way things had happened lately, that Alice had begun to think that very few things indeed were really impossible.'

Lewis Carroll

CONTENTS

CODING & CRAFT WITH THE BBC MICRO:BIT

Companion website for
micro:bit in Wonderland
alice.techagekids.com

'"Why," said the Dodo, "the best way to explain it is to do it."'

GUIDE TO THIS BOOK

CODING & CRAFT FOR THE BBC MICRO:BIT

This book guides you through twelve coding and craft projects for the BBC micro:bit. The micro:bit is a small programmable computer that allows you to explore physical computing in a fun and interactive way. The projects are designed for beginners to coding and the micro:bit. They are inspired by the story of *Alice's Adventures in Wonderland* written by Lewis Carroll.

Each chapter recreates objects and scenes from Alice's adventures providing, an imaginative backdrop for developing modern skills. Children, teens and adults will learn to code the micro:bit and make games, wearable technology, animations, music and much more.

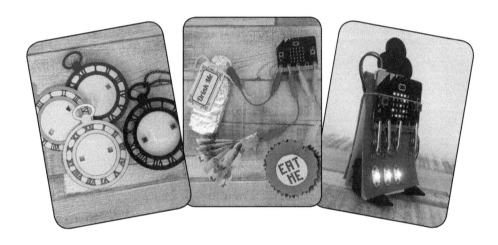

PROJECTS FOLLOWING ALICE

The projects are designed to be completed in order as you follow the story of *Alice's Adventures in Wonderland*. The story sets the stage for what you will create in each chapter. You will learn new concepts and skills with every project and build on what you have learnt as you follow one project after the other.

Reading the story of Alice before you start with a project will inspire you to be creative and experiment with the micro:bit. The original story of *Alice's Adventures in Wonderland* is available online and it's free to download and read on a device.

WHO THIS BOOK IS FOR

FOR BEGINNERS OF ALL AGES

This book is written for those who are new to coding, electronics and the micro:bit. Young learners will require adult support, however learners from around 12 years, will be able to complete the projects independently. It also provides a quirky way for older learners to have a gentle introduction to technology skills.

FOR GIRLS AND BOYS (AND ADULTS TOO)

As women in the technology industry, we wanted the projects to appeal to girls and encourage them to develop design and tech skills.

Being parents of boys, we didn't want to create a girls-only book. The projects have been tested with boys and girls and appeal to both.

Children don't divide neatly into stereotypical groups. Some girls will be fascinated by using real electronics components and some boys will love cross-stitch. Real children are a mix of skills and interests and sometimes they surprise you.

We wanted the book to be a shared experience for families—girls, boys and adults!

FOR FUTURE MAKERS

This book helps children who are drawn to computer screens to engage with technology in a creative way by exploring physical computing and developing their design and craft skills.

We also want to encourage children who enjoy art and design to add coding and electronics to their skill set.

It provides an opportunity for parents and teachers to also learn modern making and coding skills alongside their children.

WHAT YOU WILL LEARN

We hope learners and adults supporting them will discover a love of making and develop essential design and computational thinking skills.

CODING

You will learn introductory computer programming (coding) in a drag and drop editor designed for beginners. Coding concepts, such as variables, selection and loops, are introduced in a practical way.

ELECTRONICS

You will learn the basics of working with a microprocessor (mini computer) and electronics components. You'll have an opportunity to work with LEDs (lights), buzzers, circuit wiring and more.

DESIGN & MAKING

You will learn design thinking and work with a variety of tools and materials that will develop your making and construction skills. There are opportunities to use tools such as craft cutters, laser cutters and 3D printers.

HISTORY OF TECHNOLOGY

You will discover fascinating facts about the history of technology during the time period when Lewis Carroll authored *Alice's Adventures in Wonderland*.

MAKING IT YOURS

You will learn the skills needed to eventually imagine, design and make your own projects by completing the **Challenges** and **Make It Yours** sections in each chapter.

WHAT YOU WILL NEED

EQUIPMENT AND MATERIALS

You will need:

♣ a micro:bit with a USB cable and battery pack (available as a starter kit);

♣ a computer to program the micro:bit (desktop/laptop/Chromebook/Raspberry Pi/ mobile device);

♣ the makecode.microbit.org editor which is free to use and runs in a web browser (no download or installation required);

♣ electronics components (kits are available);

♣ craft materials (many of which you will already have at home or school); and

♣ a copy of *Alice's Adventures in Wonderland* (available as a free download).

Although you can do all the projects in this book with one micro:bit, it is useful to have an additional micro:bit so you don't have to take apart your favourite project immediately.

WEBSITE AND TEMPLATES

The alice.techagekids.com website accompanies this book and provides additional information, including a shopping list of equipment and materials from suggested retailers.

Templates are available on the website. Print on regular A4/letter size paper/card. When you print the templates make sure the print scale is 100% and the sheet is aligned centrally in both directions.

If you don't have access to a printer you can create your own designs.

NOTES FOR ADULTS

PARENTS

We hope that parents will enjoy sharing the story and working on the projects with their children.

We know that some parents are nervous about their lack of knowledge when it comes to technology. This is an opportunity to learn alongside your child and overcome that fear of the unknown.

TEACHERS

The projects in this book develop skills that are central to Computing and Design & Technology subjects. They can be used to deliver parts of the UK National Curriculum and the US Common Core.

The micro:bit and the additional materials used are inexpensive and readily available from educational suppliers.

The projects are ideal for cross-curricula lessons, linking with history, music, physical education, drama and, of course, English.

CLUBS, LIBRARIES AND MAKERSPACES

It's great fun to complete these projects in a workshop setting with a group of learners. The projects can be completed intensively during a holiday club or camp, or spread out over several weeks in a regular STEM/STEAM club.

The literary basis for the projects makes them a great fit for a library which is developing a makerspace or encouraging digital skills.

Makerspaces often attract adults who mix craft and technology to make theatre props or role play accessories. These projects are an engaging way to develop those interests in the next generation. Access to a laser cutter, craft cutter and 3D printer gives plenty of opportunities to extend the projects.

SHARING ON SOCIAL MEDIA

We'd love to see what you make. At the end of each chapter there's a purple heart to remind you to share your project on social media.

Remember when sharing to keep your personal information private. Take note of the age restrictions on social media platforms, children should ask a responsible adult to share their creations online.

🌐 **Find** Tech Age Kids on Facebook, Twitter and Instagram and share your projects using the hashtag *#techalice*.

'"Curiouser and curiouser!" cried Alice'

HOW TO USE THIS BOOK

Each chapter tells you to read part of *Alice's Adventures in Wonderland* and provides a materials list specifying everything you need.

Read the story, gather materials and print templates before you start working on a project.

The materials in each project are organised into an Essentials list (basics you need to complete the project) and a Get Creative list (additional items that allow you to be more creative).

Check the website for the latest tips on where to purchase electronics and craft supplies.

Use a storage box with compartments, like a sewing, craft, jewellery or tool box, to organise your resources and store accessories you've made.

ICONS IN THIS BOOK

We use icons throughout the book to highlight specific tasks or information.

📖 **Read** a chapter in *Alice's Adventures in Wonderland*.

💬 *Quotes from the story.*

🖨 **Print** a template on card or paper to make the project.

🌐 **Visit** the website alice.techagekids.com for more information.

▶ **Play** or test your code in the editor simulator.

⬇ **Download** your program from the editor and transfer it to the micro:bit.

⚠ **Warning:** Take note of a safety message.

🐛 **Troubleshoot** your code or electronics.

📑 **Tip,** hint or signpost to help you along.

∴ **Think** or challenge to extend your thinking.

> Technical terms will be defined in boxes such as this one!

'...and she set to work very carefully'.

THE BBC MICRO:BIT

MEET THE SMALL COMPUTER

The BBC micro:bit is a small, programmable computer that has built-in inputs and outputs, with the capability to connect more. You can use it to make a wearable device, cool gadgets, useful science equipment and creative craft and coding projects.

The BBC micro:bit is the result of a collaboration of over 30 partners led by the BBC (British Broadcasting Corporation).

The Micro:bit Education Foundation (microbit.org) is now responsible for the future of the micro:bit.

The micro:bit measures about 4x5cm and can be powered from a battery pack with 2 AAA batteries.

MICRO:BIT FEATURES USED IN THIS BOOK

The features of the micro:bit used in this book are:

- ♣ the display—a matrix of 25 red LEDs;
- ♣ the A and B input buttons on the front and the reset button at the back;
- ♣ the built-in accelerometer (motion sensor) and magnetometer;
- ♣ the input/output pins for connecting additional electronics components; and
- ♣ the sound output.

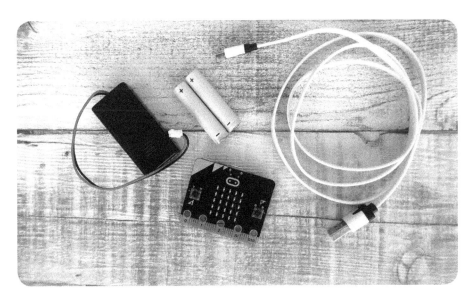

USING YOUR MICRO:BIT SAFELY

The micro:bit is designed to be used in creative projects, as with any electronic device use it with care to prevent damage and stay safe.

The projects are suitable for beginners and are intended to be completed alongside an adult or with adult supervision. The projects are undertaken at your own risk.

Read the micro:bit safety advice at microbit.org/guide/safety-advice.

Also, check alice.techagekids.com for any updated advice before you begin.

The tips below are not a substitute for reading the safety advice, however, we want to highlight some key points:

♣ Projects combine craft and tech activities. It's important that you have a tidy workspace. Make sure that materials don't unintentionally touch the micro:bit.

♣ When you are not using the micro:bit, unplug the device and put it away.

♣ Only hold the micro:bit by its edges when it's in use. You can use a case for some projects—make sure it's easy to remove.

♣ The micro:bit is designed to run cold. If yours is hot, stop using it and check the safety advice.

♣ None of the projects require you to connect crocodile clips to the micro:bit pin marked 3V (power supply pin).

PROGRAMMING THE MICRO:BIT

Find a guide on how to program the micro:bit at: microbit.org/guide/quick/.

MAKECODE EDITOR

You'll use the Microsoft MakeCode Block Editor, with drag and drop code blocks, to programme the micro:bit. The editor is free and doesn't require installing as it runs in a web browser. It runs on a Windows, Mac or Linux laptop or desktop computer. It also runs on Chromebooks, Raspberry Pi computers and Android or iOS devices.

🏮 Find the editor at makecode.microbit.org.

In the editor you'll find an on-screen micro:bit (simulator), sections containing code blocks and a coding area where you'll drag blocks to program the micro:bit.

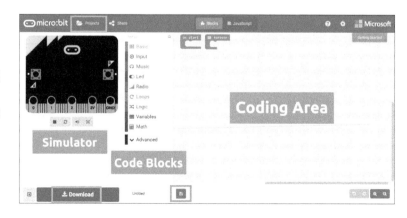

CODE BLOCKS

Code blocks are colour-coded and categorised in sections, such as Basic, Input, Loops, etc. Each code block name in this book is colour-coded to give you a clue in which section of the editor to find it.

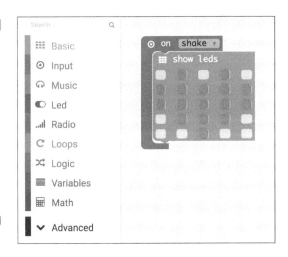

For example, show leds is a blue code block found in the Basic section. Sometimes you need to click **...More** in a section to see more blocks in that category.

If you can't find a block, use **Search** to help.

Code blocks will snap together when you drag them closer to each other. When a code block is greyed-out in the coding area, it is inactive.

SIMULATOR

The editor has an on-screen micro:bit which allows you to test your program in the browser without a real micro:bit.

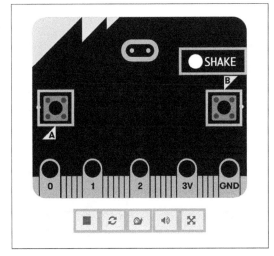

You can, for example, click button **A** and **B** and move your cursor over the on-screen micro:bit to tilt it left and right.

When you use the code block on shake **SHAKE** appears on the micro:bit in the editor. You can click the button to run the program in the simulator.

The buttons under the micro:bit will (from left to right) run your program ▶, restart your program ↻, slow everything down (snail icon), mute the sound 🔊 and show the on-screen micro:bit in full-screen mode ✛.

Some micro:bit functionality can't be tested in the simulator, so you need to learn how to download and transfer your program to your micro:bit.

SAVING AND TRANSFERRING FILES

The first project, LATE!, is specifically designed to familiarise yourself with the editor and transferring **.hex** files to your micro:bit.

> A **.hex** file stores your program and editor layout information in a format that can be read by the micro:bit and the MakeCode editor.

SAVING

Projects are automatically saved in the browser. You can also save your project by giving it a name and clicking the **save** button. Click **Download** to save your program to your computer. The *.hex* file will save to your **Downloads** folder on your computer.

You can import projects by clicking **Projects** and then **Import File** to upload a saved *.hex* file. Browse your folders to find your downloaded *.hex* file.

Important: The editor just creates a *.hex* file, it doesn't transfer it to the micro:bit!

TRANSFERRING

When using the micro:bit with a computer it will appear as a drive in your file explorer when you connect it with a data and power micro USB cable.

Transferring the *.hex* file to the micro:bit works differently with different browsers and operating systems.

You can just drag and drop the downloaded *.hex* file into the `MICROBIT` drive on your computer.

Tip: The first project walks you through transferring *.hex* files to your micro:bit using Microsoft Edge, Google Chrome and Firefox browsers.

💬 *'Alice was beginning to get very tired of sitting by her sister on the bank, and of having nothing to do ...'*

'"Oh dear! Oh dear! I shall be late!"... when the Rabbit actually took a watch out of its waistcoat-pocket, ...'

LATE!

CHAPTER 1

FOLLOW THE STORY

📑 **Read** Chapter 1 *Down the Rabbit-Hole* to meet the White Rabbit.

At the beginning of the story, Alice sees a White Rabbit with a pocket watch. The White Rabbit is most concerned about being late '*"Oh dear! Oh dear! I shall be late!".'*

Alice's Adventures in Wonderland was written in 1865 when a pocket watch would have been a pretty cool piece of technology.

YOU WILL MAKE

In this project you will program the micro:bit to be a modern version of the White Rabbit's pocket watch which always shows the same time—LATE— when you shake it.

This project will help you become familiar with the editor and transferring programs to your micro:bit.

YOU WILL LEARN

This chapter introduces **strings**, sending **output** to the micro:bit display, **repetition** and running code when an **event** occurs. It introduces working with the micro:bit including using the **accelerometer** to detect shaking, **downloading** and **transferring** programs, and **resetting** the device. The project uses paper crafting skills and the option to make use of a **craft cutter** or **laser cutter**.

(Don't worry if you don't understand terms, we'll be explaining them as you work through the project!)

YOU WILL NEED

ESSENTIALS

- ♣ micro:bit (USB cable and battery pack)

GET CREATIVE

- ♣ 2 loom bands
- ♣ Card (cream/off-white)
- ♣ Black pen, colouring pencils/felt tip pens
- ♣ Scissors, craft knife and cutting mat
- ♣ Craft cutter/laser cutter (optional)

TEMPLATES

🖨 **Print** a pocket watch template from the website on cream/off-white card/paper.

There are three different versions to choose from—print and use, print and colour-in or print and draw your own Roman numerals.

🌐 **Find** another template on the website to cut the pocket watch shape with a craft cutter or a laser cutter.

CREATE A PROJECT

Open the micro:bit editor, <u>makecode.microbit.org</u>, in a browser on your computer. The editor shows an on start and forever block in the coding area when you first open it.

CODING

SCROLLING TEXT

You're going to create your first **string** for the micro:bit.

> In coding, a **string** is a sequence of characters such as the ones you type on a keyboard.

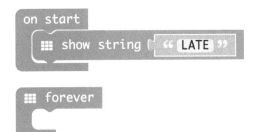

> Drag a show string block from the Basic section into the on start block and edit the text so it says LATE.

☷ **Tip:** Right-click in the coding area and select **Format Code** to organise your code neatly.

Placing the show string block inside the on start block will scroll the text—LATE— once when you first run your program. It runs automatically in the **simulator**.

> The **simulator** allows you to try out your program on the computer screen without a real micro:bit. It's useful for testing your programs.

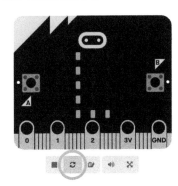

You can rerun your code by pressing the **refresh icon** under the on-screen micro:bit.

⊙ **Try** run your code more than once in the simulator.

TRANSFER THE PROGRAM TO YOUR MICRO:BIT

Now you need to download and transfer your program to your micro:bit. Transferring the *.hex* file to the micro:bit is simple, but works differently with different browsers and operating systems. We're going to walk you through two methods, using Microsoft Edge and using Google Chrome with the micro:bit *Uploader* program.

SAVE YOUR PROJECT

First, give your project a name. In this case, call it LATE.

🚏 **Note:** Your project will also be saved automatically in the browser. You can click **Projects** to find previously saved programs in your browser. If you haven't given your project a name it will be saved as *Untitled.hex*.

CONNECT YOUR MICRO:BIT

Connect the micro:bit to your computer using a micro USB cable. You should see a yellow power light come on at the back of the micro:bit.

The USB cable allows you to power the micro:bit with your computer and transfer your program to the micro:bit from the editor.

A long USB cable makes it easier to move the micro:bit around, whilst being powered by your computer.

If you want to move the micro:bit away from the computer and run your program you need to power it with a battery pack.

TRANSFER USING MICROSOFT EDGE

In Microsoft Edge and Internet Explorer, click **Download**. The save prompt will appear at the bottom of the browser. Choose **Save As** from the save prompt.

Now, in your file explorer, select the **MICROBIT** drive as the location to save the *.hex* file. Click **Save**.

🔔 **Remember:** Click **Save As** in the save prompt and not **Save** or your program won't be transferred.

TRANSFER USING GOOGLE CHROME WITH THE UPLOADER

If you are using a Chrome or Firefox browser then the micro:bit *Uploader* program is really useful. The *Uploader* is available as an experimental free download for Windows that makes transferring *.hex* files a lot easier.

Go to alice.techagekids.com and click *Uploader* to access the program.

You need to run the *Uploader* once at the beginning of each coding session, then when you click **Download** in the editor it will copy the downloaded *.hex* file to the micro:bit automatically.

This method makes downloading your program to the micro:bit a one-click process.

Tip: Remember to run the *Uploader* at the start of a coding session and use the standard downloads location in your browser for this to work.

OTHER PLATFORMS

On Mac, Chromebook, Linux and Raspberry Pi, drag your *.hex* file from the **Downloads** folder to the MICROBIT drive.

There are also Android and iOS mobile apps for flashing (transferring) *.hex* to the micro:bit wirelessly.

TRANSFER YOUR PROGRAM

Download and transfer your program to your micro:bit.

The yellow light will flash on the back of the micro:bit when your program is being transferred.

When the light stops flashing your program will run once. To re-run the program press the reset button on the back of the micro:bit.

Tip: Always check that the light is flashing when you transfer a program from a computer.

FOREVER LATE!

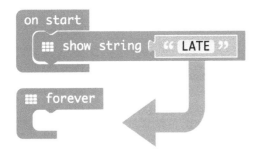

Now *change* your code by dragging the show string block inside the forever block.

Do this before deleting the on start block.

If you make a mistake you can undo your last action by typing **Ctrl Z/cmd Z** on the keyboard or clicking the buttons in the editor. You can also **zoom in** and **out** to make your code bigger or smaller in the coding area.

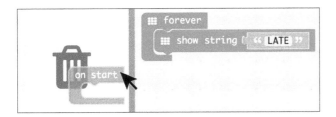

Drag the on start block off the code area to delete it. A bin will appear to delete the block. You can also tap **delete** on the keyboard.

Now you should see your string scroll on the simulator continuously. The forever block creates a **repetition** that never stops.

> **Repetition** is running a sequence of instructions multiple times.

🏳 **Tip:** If you ever delete an on start or forever block and then find that you need one, they are in the Basic section.

⬇ **Download** and transfer your program to the micro:bit. Remember clicking **Download** only saves the file. Unless you are running the *Uploader* you still need to transfer the file to your micro:bit.

Practise transferring your program to learn how it works with your setup.

SHAKE ACTIVATION

You don't need to show the string LATE all of the time. The micro:bit has an **accelerometer** (movement sensor) and it can detect when you shake it.

> An **accelerometer** measures changes
> in speed and direction.

Add the on shake block from the Input section to your coding area. Code blocks inside the on shake block will run every time you shake the micro:bit.

Change your code again by dragging the show string block into the on shake block. Delete the forever block.

Now you should see a **SHAKE** button appear on the simulator.

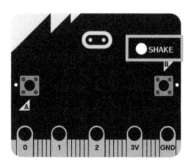

Click the **SHAKE** button on the simulator to see LATE scroll on the display.

The micro:bit can run code when an input **event**, such as shaking, is detected.

> An **event** is something that a program can
> react to. The micro:bit generates events
> when the accelerometer detects shake,
> tilt and other movement patterns.

TEST YOUR CODE

⬇ **Download** your program and transfer it to your micro:bit.

You can test your code whilst the micro:bit is still attached to the computer with the USB cable.

Shake the micro:bit to see LATE scroll across the display. Practice moving the micro:bit to see what movement triggers the on shake event.

ATTACH THE BATTERY PACK

If you have a battery pack then you can disconnect the micro:bit from the computer and power it from the battery pack.

⚠ **Carefully** attach or remove the connector on the battery pack so that you don't pull the wires out of the connector. You may need an adult to help as the connection is quite fiddly.

Use two loom bands to secure the battery pack to the micro:bit. Place the battery pack on the back of the micro:bit and stretch the loom bands over the micro:bit and battery pack as shown below.

 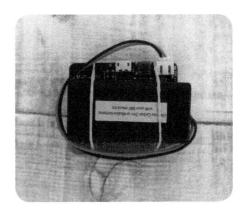

Loom bands provide a great temporary way of attaching the battery pack to the micro:bit. Practice this technique as you'll be using it throughout the book!

⚠ **Warning:** Cases and attachments that touch the micro:bit need to be non-conductive. You'll learn more about what this means in the next chapter.

A BIT OF HISTORY

POCKET WATCH

Alice's Adventures in Wonderland was published in 1865. This was an important time in the history of the pocket watch. In 1857 the American Watch Company made the first pocket watches from standard parts. Before that each watch was individually made and repairing them was expensive.

MAKING A WATCH FACE

It's time to add a watch face to your micro:bit so that it looks like the kind of old-fashioned pocket watch that the White Rabbit would have used.

CARD POCKET WATCH

The instructions how to make the pocket watch are on the template.

If you printed the template on paper stick it to card to make the pocket watch more sturdy.

Gently press the card watch face over the front of the micro:bit so that the buttons come through the holes you have created. The watch face will be held in place by the buttons and can easily be removed. No glue needed!

If your button holes end up too big, use some sticky tack on the top corners of the micro:bit to hold the watch face in place.

ROMAN NUMERALS

Roman numerals were traditionally used on pocket watches and still appear on some watches and clocks today.

In Roman numerals, I is 1, II is 2 and III is 3. V is five. To get 4 place an I before a V to give IV (if a smaller number appears before a larger one, you take it away.) VI is 6, VII, is 7, VIII is 8.

X is ten. To get 9 place an I in front of X to give IX (10 take away 1). XI is eleven and XII is twelve.

So the Roman numerals you need on a watch face are:

I, II, III, IV, V, VI, VII, VIII, IX, X, XI and XII.

CRAFT CUTTER / LASER PRINTER POCKET WATCH

The intricate design of the pocket watch makes it a perfect project to try out using a craft cutter or laser cutter.

The pocket watch outline on the left was cut with a **craft cutter** on black card. The one on the right was cut using a **laser cutter** and black acrylic.

In both cases you need to add a card insert to secure the micro:bit. (Instructions are available on the website.)

> **Craft cutters** use a blade to cut a design out of materials such as paper, card or vinyl.
>
> **Laser cutters** use a laser to cut a design out of materials such as wood or acrylic.

💬 *'...burning with curiosity, she ran across the field after it [rabbit], and fortunately was just in time to see it pop down a large rabbit-hole under the hedge.'*

Share your make #techalice

MAKE IT YOURS

Customise your project in your own style:

♣ See what other inputs you can use to trigger the text.

♣ Change the text that scrolls on the display.

♣ Decorate the pocket watch accessory to suit your style.

'"What a curious feeling!" said Alice, "I must be shutting up like a telescope!"'

DRINK ME, EAT ME

CHAPTER 2

FOLLOW THE STORY

📖 **Read** Chapter 2 *The Pool of Tears.* You've already read Chapter 1 *Down the Rabbit-Hole*. In these chapters you discover what makes Alice change in height.

A lice finds several things in Wonderland that make her grow or shrink. At the start of the story she finds herself in a room with a small table. On the table are a key and a bottle and underneath it is a cake in a little glass box.

YOU WILL MAKE

In this project, you're going to make the display on the micro:bit show Alice's size when she drinks from the bottle labelled *Drink Me* and eats the cake marked *Eat Me*.

You'll be making external inputs for the micro:bit using conductive materials.

The restricted display panel on the micro:bit requires you to apply your imagination for this project!

YOU WILL LEARN

This chapter explains how to use a conductive object with the micro:bit to create an external **input**. It involves connecting **crocodile clip leads** to **pins** on the micro:bit. It introduces **selection** (if/then conditional blocks.) There's opportunity to develop practical **craft skills** with paper, card, foil and play dough.

YOU WILL NEED

ESSENTIALS

- ♣ micro:bit (battery pack and USB cable)
- ♣ 4 crocodile (alligator) clip leads (with sprung metal clips on each end to connect things temporarily)
- ♣ Paper (plain paper or decorative paper)
- ♣ Kitchen foil
- ♣ Foil cake case (or make one using kitchen foil)
- ♣ Conductive coin (two pence piece)
- ♣ Hole punch, craft glue, sticky tape, scissors, string, black pen

GET CREATIVE

- ♣ Card
- ♣ Decorative cupcake case
- ♣ Air drying clay (black to make the currants)
- ♣ Play dough (recipe available on the website to make your own)

TEMPLATES

🖨 **Print** the template from the website for the bottle shape, bottle label and cake topper on paper/card.

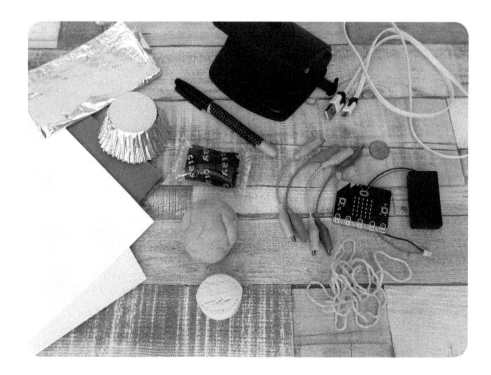

START A NEW PROJECT

Click on **Projects** and select **New Project** to create a new project. The coding area will reset showing only the on start and forever blocks.

🔖 **Tip:** If you forgot to save your previous project it should have saved automatically in the browser. Remember to give your projects a name, so you can find them again another time.

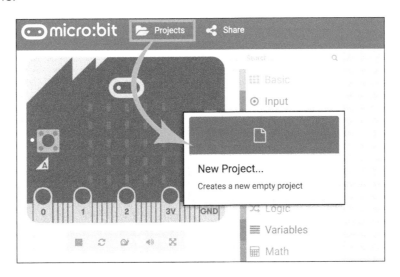

CODING

SHOW ALICE

Imagine that Alice, at her normal height, is 3 pixels tall on the micro:bit.

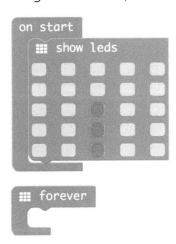

> *Drag* the show leds block from the Basic section into the on start block and click the middle 3 pixels from the bottom to turn them red.

🔖 **Tip:** One click will turn the LED red and if you click the same LED again it will clear it.

DRINK ME CODE

The micro:bit has three large **pins** labelled 0, 1 and 2 that you can easily connect to. The micro:bit can tell when there is an electrical connection between its GND (ground) pin and pins 0, 1 or 2. The 3V pin is a power supply pin and not used in this project.

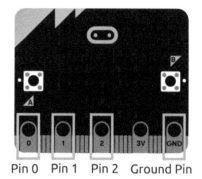

Pin 0   Pin 1   Pin 2   Ground Pin

> A **pin** carries electrical signals in and out of a computer.

In the story, when Alice drinks from the bottle she shrinks to 10 inches (about 25cm) tall.

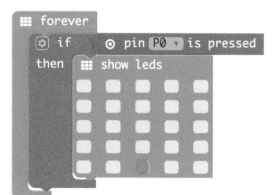

Add this code to the forever block to make Alice shrink when a connection is made to Pin 0.

The if...then block is in the Logic section and the pin is pressed block is in the Input section.

If Pin 0 is pressed (you're making an electrical connection between Pin 0 and Ground) the micro:bit should display a 1 pixel tall Alice.

> **If** blocks allow you to do something only if a **condition** is true. This is called **selection**.

Test the code by clicking on Pin 0 in the simulator. Did you notice what happened? You can click the **reset** button to see it again.

 Download your program to the micro:bit.

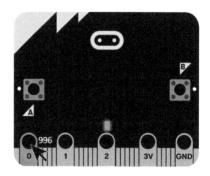

DRINK ME BOTTLE

💬 *'... and round the neck of the bottle was a paper label, with the words 'DRINK ME' beautifully printed on it in large letters.'*

Your bottle needs to be **conductive**. Use the template of the bottle shape and make a conductive bottle from layers of packaging card glued together and then wrapping it in kitchen foil.

> **Conductive materials** allow electric current to flow through them. They can be used in an electrical circuit.

Now cut out the *Drink Me* label from the template or make your own. Make a hole with the hole punch and attach the label to the bottle with some string.

CONNECTING THE BOTTLE

You'll use crocodile clip leads to connect conductive objects to the micro:bit as **inputs**.

> The micro:bit has a range of **inputs** that allow it to detect changes in the physical world. These include built-in buttons and sensors as well as additional input devices that can be connected to the pins.

Take a crocodile clip lead and connect one end to a conductive coin or another conductive metal object (this just gives you a bigger surface to touch.) Connect the crocodile clip on the other end of the wire to GND.

Take another crocodile clip lead and connect one end to your bottle. Connect the other end to Pin 0.

⚠ **Carefully** connect the crocodile clip leads so that they are straight on and don't accidentally touch other pins.

The thin gold strips in-between the five large pins are smaller pins which you should also avoid accidentally connecting to.

We won't be using the smaller pins in this book!

TESTING THE BOTTLE

Now touch the coin with one hand and then pick up the bottle with your other hand. You've completed an **electrical circuit** and Alice should shrink to one pixel!

> An **electric circuit** is a loop that allows electrical current to flow from a power source, through components and back to the power source.

You don't have any code to make Alice grow again (yet!) so she will stay small until you reset the micro:bit.

🖥 **Tip:** Press the **reset** button on the back of the micro:bit to try again.

⚠ **Warning:** The electrical current here is very small so it's not dangerous and the micro:bit has been designed to be used in this way. Never, ever experiment with electricity from mains sockets.

🐞 **Troubleshoot:** Crocodile clips have an insulating sleeve to stop them accidentally making connections. The sleeve sometimes slips off and can be put back on by an adult. Push the sleeve back a little when you need to hold it or push it into play dough to expose more of the metal.

CODING

EAT ME CODE

When Alice eats the cake she grows to 9 feet (about 275cm) tall!

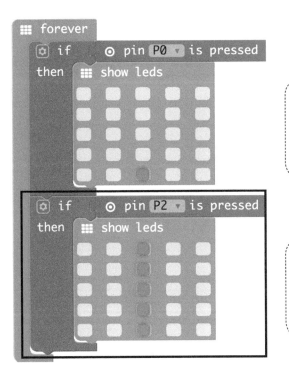

Tip: You can right-click on any code block to create a duplicate of that block. Try it with the logic block, if..then, when you add the code below.

Add more code inside your forever block to detect the cake being touched.

Did you notice you used Pin 2 rather than Pin 1? You'll use Pin 1 later!

▶ **Test** your code in the simulator. You should be able to alternate between clicking **Pin 0** and **Pin 2** to shrink and grow Alice.

⬇ **Download** your code to the micro:bit.

EAT ME CAKE

💬 *'Soon her eye fell on a little glass box that was lying under the table: she opened it, and found in it a very small cake, on which the words 'EAT ME' were beautifully marked in currants.'*

> **Currants** are small dried grapes. Nothing to do with electrical currents!

You can use anything conductive to make the cake. You could even use a real cake, but don't eat it afterwards! (Crocodile clips are not food-safe.) We're going to show you two methods—a paper cake with a foil cupcake case and a play dough cake.

PAPER CAKE

Make a cake from paper and then put it in a conductive foil cupcake case. Write EAT ME in the centre of a piece of paper so that it looks like it was made by placing currants. The total width of your writing should be about 3-4cm.

Crumple up the piece of paper keeping the EAT ME visible on the top. Use sticky tape or craft glue to stick the paper cake into a foil cupcake case. You can also make a conductive cupcake case using kitchen foil.

PLAY DOUGH CAKE

Make a cake from play dough and put it in a cupcake case. You can use regular play dough or make your own. Cut out and place the cupcake topper (from the template you printed earlier) onto your cake. You can also write EAT ME with tiny balls made using black air drying clay.

CONNECTING THE CAKE

Now connect a crocodile clip lead from the conductive cake (or foil cupcake case) to Pin 2 on the micro:bit.

⚠ **Carefully** connect the crocodile clips as shown.

⚠ **If** you're using a play dough cake, be careful not to allow play dough on or near the micro:bit. You may also need to pull the protective sleeve back to expose more of the metal on the crocodile clip and push it into the play dough.

TESTING THE CAKE

Now, touch the coin with one hand and touch the cake with your other hand to make Alice grow. Remember, paper isn't conductive so you need to touch the play dough or foil case depending on how you made your cake. Touch the bottle and cake and make Alice shrink and grow.

🐞 **Troubleshoot:** Make sure your bottle and cake don't touch.

⌨ **Hint:** Don't forget that you must continue to touch the coin with one hand to complete the circuit.

⚠ **Clean** and dry the crocodile clips well after using them with play dough!

💬 *'Alice took up the fan and gloves, and, as the hall was very hot, she kept fanning herself all the time she went on talking: "Dear, dear! How queer everything is to-day! And yesterday things went on just as usual. I wonder if I've been changed in the night?"...'*

CHALLENGE

MAKE ALICE 2 FEET TALL

When Alice picks up the White Rabbit's fan she starts shrinking until she realises she is 2 feet (61cm) tall.

Can you make a fan that shrinks Alice to 2 feet tall when you touch it?

Make a paper concertina fan with tin foil covering the bottom part.

🔀 **Hint:** You need to add a code block for P1 and connect the fan to Pin 1 on the micro:bit using a crocodile clip lead.

⚠ **Carefully** connect the crocodile clips as shown.

Now touch the bottle, cake and fan alternately and watch Alice change in size.

🔀 **Remember** to name your project and save it in the editor.

💬 *'She ate a little bit, and said anxiously to herself, "Which way? Which way?", holding her hand on the top of her head to feel which way it was growing, and she was quite surprised to find that she remained the same size: to be sure, this generally happens when one eats cake, but Alice had got so much into the way of expecting nothing but out-of-the-way things to happen, that it seemed quite dull and stupid for life to go on in the common way.'*

MAKE IT YOURS

Now it's time to customise the project.

♣ Get creative and design the fan, bottle and cake in your own style. Remember they all need to include something conductive.

♣ Improve the animation of Alice on the micro:bit.

♣ Try using other objects to control an animation on the micro:bit. You can only connect three items at a time in this way but you can swap them.

'At last the Dodo said, "Everyone
has won, and all must have prizes."'

THE CAUCUS RACE

CHAPTER 3

FOLLOW THE STORY

📖 **Read** Chapter 3 *The Caucus-Race and a Long Tale* to find out about the race.

💬 *'There was no 'One, two, three, and away,' but they began running when they liked, and left off when they liked, so that it was not easy to know when the race was over.'*

In the Caucus Race all the creatures run round and round in circles until the Dodo stops the race and declares that everyone has won. A caucus is a kind of meeting (often in politics). They don't always make sense.

YOU WILL MAKE

In this project you will create a pedometer to count your steps when you run in your own race.

You will make a strap for your shoe, so that you can wear your micro:bit.

YOU WILL LEARN

This chapter introduces the concept of storing data in a **variable.** It also shows how a simple **pedometer** works and explores the design of a practical **wearable** device.

YOU WILL NEED

ESSENTIALS

- ♣ micro:bit (USB cable and battery pack)
- ♣ Elastic band (or make a chain with loom bands)
- ♣ Card
- ♣ Colouring felt tip pens/colouring pencils
- ♣ Scissors (or a craft knife and cutting mat)

GET CREATIVE

- ♣ MI:power board from Kitronik
- ♣ Duct tape, loom bands
- ♣ Hole punch
- ♣ Prizes (In the story comfits and a thimble are used as prizes.)

TEMPLATE

🖶 **Print** the template for a wearable from the website on card, preferably, or paper.

🌐 **Find** a template to cut the wearable shape with a craft cutter on the website.

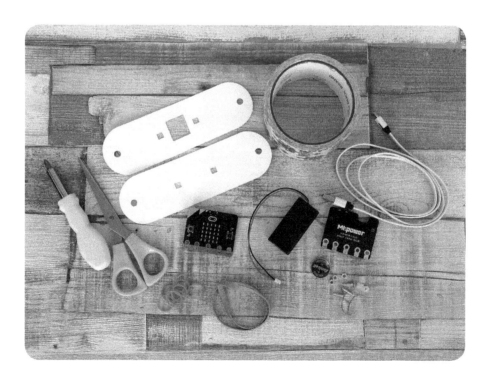

USING A *STEPS* VARIABLE

A pedometer keeps track of how many steps you've taken. You need to use a **variable** to keep track of the steps.

> In coding a **variable** is used to store changeable values that you need to refer to in a program.

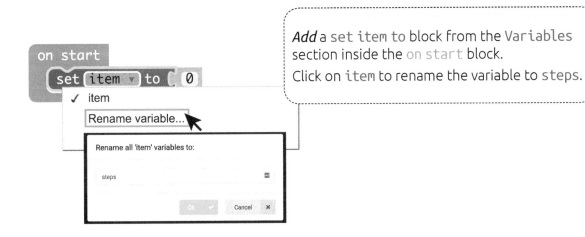

Add a set item to block from the Variables section inside the on start block.

Click on item to rename the variable to steps.

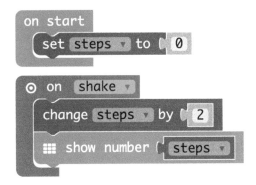

Add these code blocks so that every time the micro:bit detects a shake you increase the value of the steps variable by 2. This is because the micro:bit will be worn on one foot and won't detect steps taken by the other foot.

It's also useful to be able to see how many steps have been detected, so show that too.

▶ **Test** your code in the simulator by clicking the **SHAKE** button.

⋰ **Think:** What happens to the number on the display when it becomes bigger than 9?

TEST THE PEDOMETER

It's time to test out your pedometer.

⤓ **Download** your program to the micro:bit.

Disconnect the micro:bit from the computer and attach the battery pack. Use an elastic band to tie the micro:bit around your shoe. Tuck the battery pack into your sock or secure it to the micro:bit with a couple loom bands as you did in the first chapter. You'll have an opportunity to make a better wearable later.

❧ **Hint:** To restart the step count, press the reset button on the back of the micro:bit.

Now count your steps as you walk. Check if it matches the number shown on the micro:bit.

ENDING THE RACE

At the end of the Caucus Race everyone receives a prize. Alice has some comfits, which are old-fashioned sweets, in her pocket and gives them out. Alice receives her own thimble as a prize.

Let's display a comfit image on the screen when you reach your target. Start with a small step target (10) so that you can test your program using the simulator.

```
on start
  set steps ▾ to ( 0 )

● on  shake ▾
  change steps ▾ by ( 2 )
  ▦ show number ( steps ▾ )

▦ forever
  ⚙ if ( ( steps ▾ ) ≥ ▾ ( 10 ) )
  then  ▦ show leds
```

> *Add* two logic blocks if..then and >= (into the forever block) to show an image of the comfit on the display when you've taken 10 or more steps.

▶ **Test** your code in the simulator by clicking the **SHAKE** button until you see the comfit.

Once you're happy that the program is working you can increase your target number of steps. Perhaps agree a small prize (as in the story) when you reach your target.

⤓ **Download** your code to the micro:bit, detach from the computer and attach the battery pack or a MI:power board.

MI:POWER BOARD

The micro:bit can also be powered by a coin cell battery using the MI:power board, from Kitronik. This allows you to attach it to your arm or belt or hang it around your neck, without worrying about a bulky battery pack. Visit www.kitronik.co.uk for more information about the MI:power board and how to attach it to your micro:bit.

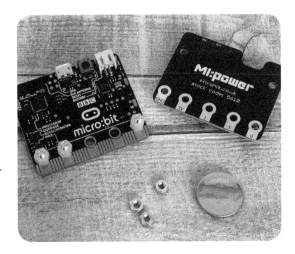

The MI:power board also has a built in buzzer which will be useful in other projects.

⊟ **Note:** If you wear the micro:bit around your neck then you only need to increase the step count by one on each shake.

CRAFT

MAKE A WEARABLE

You can now make a **wearable** from card or duct tape to fit the micro:bit over your shoe.

> A **wearable device** is an electronic device that can be worn as a smart watch or accessory and often uses sensors to collect data.

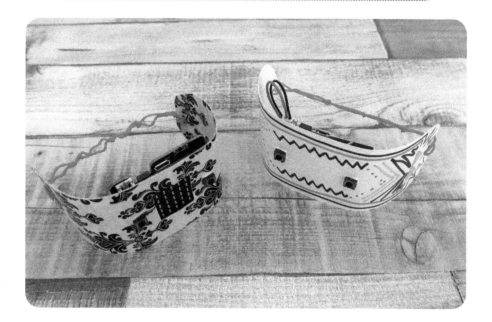

PAPER WEARABLE

If you printed the template on paper, stick it to card to make it sturdier.

Cut out the wearable shape. Cut holes for the micro:bit buttons and LED display using scissors or a craft knife. Punch holes on either end of the wearable and attach a large elastic band as shown below.

🏷 **Tip:** If you don't have a large elastic band that will fit around your shoe, make a chain-link using loom bands. You can use a *S-connector* to attach the chain of loom bands to the wearable.

Press the wearable over the micro:bit so the buttons come through the holes you made. Attach the wearable over your shoe with the elastic band.

If you haven't attached the battery pack to the micro:bit using two loom bands, you can tuck it into your sock or tie it between your shoelaces.

DUCT TAPE WEARABLE

Make a more durable wearable using a piece of duct tape. You'll need a length of duct tape twice the length of the template. Fold it back on itself so that you have the duct tape pattern on the top and bottom. Use the template to cut the holes for the micro:bit buttons and LED display. Create holes on either end of the wearable and attach the elastic band/chain of loom bands and secure it to your shoe.

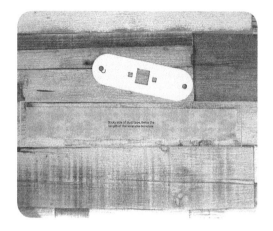

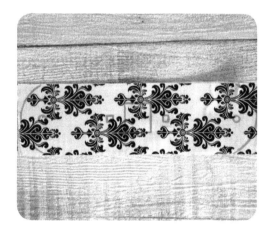

⚠ **Warning:** Don't use your micro:bit in the rain!

⧉ **Remember** to name your project and save it before moving on to another project.

"In that case," said the Dodo solemnly, rising to its feet, "I move that the meeting adjourn, for the immediate adoption of more energetic remedies—"

MAKE IT YOURS

Share your make #techalice

Now it's time to customise your step counter.

♣ Wear the micro:bit in different places (pocket, sock, wrist, top of a trainer) and see how it affects the step count. Find out what works best.

♣ Design the paper wearable to suit your style.

♣ Make the wearable from other materials, such as fabric/felt or improve on its design based on how you power the micro:bit and how you want to wear it.

'"Who are You?" said the Caterpillar.'

LIGHT UP CATERPILLAR

CHAPTER 4

FOLLOW THE STORY

📄 **Read** Chapter 5 *Advice from a Caterpillar* where Alice tries to work out who she is and what size she should be. Alice first meets the Caterpillar at the end of Chapter 4 *The Rabbit Sends in a Little Bill*.

Alice meets a Caterpillar who asks her all sorts of questions. *'There was a large mushroom growing near her, about the same height as herself. She stretched herself up on tiptoe and peeped over the edge and her eyes immediately met those of a large blue caterpillar, ...'*

YOU WILL MAKE

In this project, you will use a mushroom (real or made from play dough) to create an electrical circuit to light up the Caterpillar with an LED. When you touch both halves of the mushroom, the Caterpillar's LED will flip (toggle) between on and off.

There are different ways to complete this project depending on the materials and tools you have available.

YOU WILL LEARN

This chapter introduces working with hardware **outputs**, specifically an **LED** (light emitting diode). It introduces **Boolean** (true/false) variables and explains how to code a **toggle (on/off) switch** using a single hardware input. There are plenty of opportunities to be creative with card, paper and play dough. **3D printing** can also be used.

YOU WILL NEED

ESSENTIAL

- ♣ micro:bit (USB cable and battery pack)
- ♣ 4 crocodile clip leads
- ♣ 1 LED/sewable LED (blue or white LED)
- ♣ Paper (white if you're using a blue LED/otherwise blue paper)
- ♣ LEGO® (or other non-conductive material)
- ♣ Kitchen foil
- ♣ Scissors, glue, sticky tape
- ♣ Play dough (recipe available on the website to make your own)

GET CREATIVE

- ♣ 3D printer (clear filament)
- ♣ Air drying clay (blue)
- ♣ Silicone caterpillar mould, hot glue gun
- ♣ Real mushroom
- ♣ Conductive coin (optional)

TEMPLATES

🖨 **Print** a template from the website to make a paper Caterpillar. The template also includes instructions on how to make a mushroom from sheet plastic.

ABOUT LED LIGHTS AND RESISTORS

An LED is a diode, that means that it only works in one direction. Some LEDs are marked with + and – so that you can tell which way round they need to go. Other LEDs have one leg longer than the other, the longer leg is the positive one.

You need to insert the LED the right way round in your caterpillar.

The Caterpillar is described as being blue, so a blue LED is a good choice but you can just use a white LED. Normally, when you use an LED, you need a **resistor**. Some LEDs labelled for use in 3V e-textiles projects can be used without a resistor which are easier for beginners.

> A **resistor** reduces the flow of electrical current so that an LED doesn't get damaged.

TRY YOUR LED

The micro:bit can act as a battery and a switch. You can open and close the switch with code by writing a 0 or a 1 to a pin on the micro:bit.

First, let's try turning your LED on and off using the **A** and **B** buttons on the micro:bit.

⚠ **Warning:** In this project, we used an LED recommended for use in 3V e-textiles projects, so we didn't add a resistor. Check whether your LED requires a resistor, some kits include LEDs and matching resistors.

CODE YOUR LED

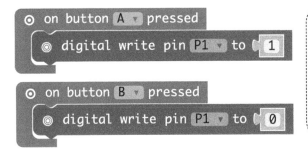

Add this code to provide power to the LED when button **A** is pressed and remove power when button **B** is pressed using the digital write pin block.

Note: Pins blocks are in the Advanced section.

▶ **Test** the buttons in the simulator. When you click on button **A** Pin 1 will be highlighted and when you click on button **B** it will turn off.

⬇ **Download** the code to your micro:bit to test your LED.

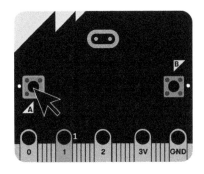

CONNECT YOUR LED

Use a crocodile clip lead to connect Pin 1 to the positive (longer) leg of your LED and another crocodile clip lead to connect GND to the negative (shorter) leg of the LED as shown in the diagram.

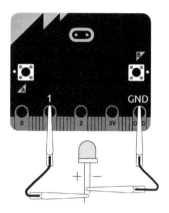  

TEST YOUR LED

Press button **A** to switch the LED on and press button **B** to switch the LED off.

🐛 **Troubleshooting:** Make sure your micro:bit is connected to the computer or attach the battery pack to power the LED.

TOGGLE SWITCH AND BOOLEAN VARIABLES

The mushroom is going to become a **toggle switch** to light up the Caterpillar's LED.

> A **toggle** switch flips between two states: on and off. The Caps Lock key on your keyboard is an example of a toggle.

When you connect both sides of the mushroom, by touching them, you will turn the LED *on* if it's off and *off* if it's on. For this, you'll use a **Boolean** variable.

> A **Boolean** variable has two states - true or false.

A BIT OF HISTORY

ABOUT GEORGE BOOLE

George Boole published a paper on Boolean logic in 1854. Alice was published in 1865. Lewis Carroll was a pen name (a name authors use for writing) for Charles Dodgson who was a mathematician and was likely to have been aware of Boole's work. Dodgson also completed important work on logic.

CRAFT

MAKING THE CATERPILLAR

Let's first make the Caterpillar which will house the LED. You can make the Caterpillar using paper, 3D printing, hot glue with a caterpillar silicone mould or design your own. The Caterpillar needs to diffuse (spread out) the light from the LED.

PAPERCRAFT CATERPILLAR

Find instructions on how to make the caterpillar on the template.

The legs of the LED will make the antennae and should be accessible to connect crocodile clip leads. The longer leg of the LED should be the Caterpillar's right antenna, so on the left as you look at it.

If you use a sewable LED then you can make antennae from short lengths of bendable wire. Make the positive antenna slightly longer so that you can tell the difference.

🚏 Tip: Make sure you've put the LED in before sticking the head on the caterpillar.

3D PRINTED CATERPILLAR

🌐 **Find** instructions on how to 3D print a Caterpillar using clear filament on the website.

Put the LED inside the head of your Caterpillar so that the legs stick out from the top.

HOT GLUE CATERPILLAR

Buy a Caterpillar-shaped silicon mould. Fill the mould with hot glue and place the LED inside the glue before it sets. Make sure the legs of the LED are free to connect crocodile clip leads.

⚠ **Warning:** Hot glue can burn. Ask an adult to help.

DESIGN YOUR OWN

Use the handle of a clear plastic milk bottle for the body and pipe cleaners for the legs. Attach the LED and make sure the legs are free to connect to.

CODING

You don't need the code you used to test your LED. Delete the button code and start a `New Project` to make the toggle switch.

You're going to use a Boolean variable called on to keep track of whether the LED is on or off. The variable on will be set to ***true*** when the LED is on and ***false*** when it's off.

```
on start
    set on to false
```

Add this code to the on start block. The false block is in the Logic section.

Think: If the variable is ***not true*** then it's false. If the variable is ***not false*** then it must be true. Boolean variables are very useful in computing.

You can use not to flip the value of a Boolean variable. If it was true then it will become false and if it was false it will become true.

Each time the mushroom switch is connected you want to flip the state of the LED.

```
on pin P0 pressed
    set on to not on
```

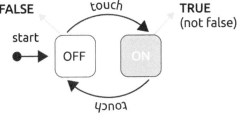

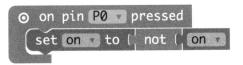

Add an on pin pressed block to detect when the mushroom switch is activated by connecting Pin 0 to GND.

The on pin P0 pressed block runs the code inside it when Pin 0 is connected and then released. You need to turn the LED on or off depending on whether on is ***true*** or ***false***.

```
on start
    set on to false

on pin P0 pressed
    set on to not on
    if on
    then
        digital write pin P1 to 1
    else
        digital write pin P1 to 0
```

Add an if..then..else logic block and set P1 to ***on*** (1) and ***off*** (0).

Tip: if on is the same as if on equals true for Boolean variables.

Download your program to your micro:bit.

CONNECTING

Use 2 crocodile clip leads to connect the LED to the micro:bit. Then connect a crocodile clip lead to Pin 0 and another to GND as shown in the diagram below.

📶 **Note:** The positive leg of the LED connects to Pin 1 and the negative side connects to GND.

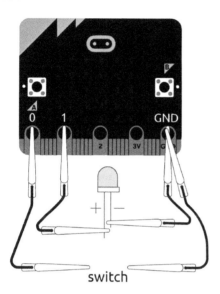

switch

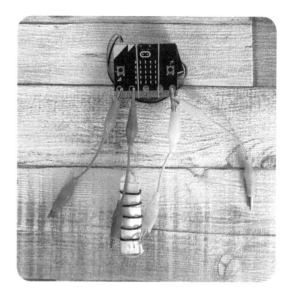

Now hold the crocodile clip from the GND pin in your hand. You may find it easier to attach a conductive coin to hold. Tap the other end of crocodile clip lead attached to Pin 0 once to turn the LED on and tap again to turn it off.

🐞 **Troubleshoot:** Check you've connected the LED legs the right way around.

MAKING A MUSHROOM

💬 *Alice remained looking thoughtfully at the mushroom for a minute, trying to make out which were the two sides of it;...*

You need to make a mushroom with two halves with an **insulator** between the two sections. You can make the insulator from modelling clay, air drying clay, plastic or another insulating material.

You must make sure that the two halves of the mushroom are completely separated. The mushroom will act as a toggle switch to turn the Caterpillar light on and off.

> An **insulator** doesn't allow electrical current to flow through it.

LEGO MUSHROOM

Design and build a mushroom shape using LEGO or any other plastic building bricks. You need to build a separator in the middle which will become your insulator.

3D PRINTED MUSHROOM

You can find instructions on the website to 3D print a mushroom shape.

REAL MUSHROOM

If you're using a real mushroom cut it in half with a knife.

DESIGN YOUR OWN

Use the template and construct a mushroom using sheet plastic, such as a plastic milk bottle. You could use air drying clay, which is non conductive, to mould the mushroom and make it how you like it.

ADD THE PLAY DOUGH

You can use play dough as the conductive material to complete the circuit. You need to add one blob of play dough on either side of the insulating material. If you don't have play dough, you could easily make some or use some kitchen foil. Your mushroom is now ready for connecting.

🚏 **Tip:** Make sure the play dough from the two sides are not touching at all.

CONNECTING

Press the other end of the crocodile clip connected to P0 into one half of the play dough (or mushroom) and the crocodile clip connected to GND into the other half of the play dough (or mushroom) as shown below.

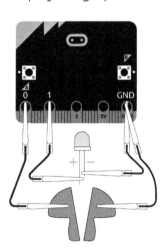
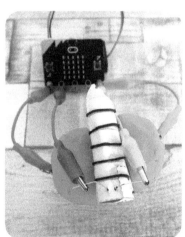

💬 *'She stretched herself up on tiptoe, and peeped over the edge of the mushroom, and her eyes immediately met those of a large caterpillar, that was sitting on the top with its arms folded,...'*

TRY IT OUT

Now that you've made all the connections, try to sit the Caterpillar on top of the mushroom, as in the story.

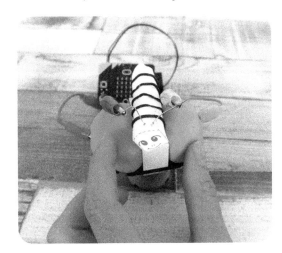 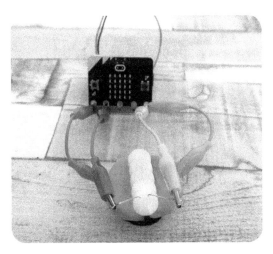

To activate the Caterpillar's LED, touch one-half of the mushroom with your thumb and the other half with your index finger *at the same time* and then move your hand away. By touching both halves of the mushroom you complete a circuit and the LED will light up (turn the LED on). Touch both halves of the mushroom again to turn the LED off.

Tip: If the LED doesn't light up try touching for a little longer so the micro:bit has a chance to detect it.

Troubleshoot: Make sure the play dough or two halves of the real mushroom are not touching at all. Check all your connections.

MAKE IT YOURS

Share your make #techalice

Now you can try this project using different materials.

♣ Can you make a Caterpillar out of materials other than paper? How about an opaque plastic such as a milk bottle?

♣ Can you create a 3D model of a Caterpillar?

♣ What other conductive material can you use to make the mushroom?

♣ Can you combine what you learnt in this chapter and the Drink Me, Eat Me chapter to make Alice grow when you touch one-half of the mushroom and shrink when you touch the other?

'"There's certainly too much pepper in that soup!" Alice said to herself, as well as she could for sneezing.'

THE COOK AND THE PEPPER BOX

CHAPTER 5

FOLLOW THE STORY

📖 **Read** Chapter 6 *Pig and Pepper* to meet the Duchess and the Cook with her peppery soup.

The Duchess's cook uses too much pepper in the soup! She carries a pepper box with her and makes everyone, apart from herself and a large cat, sneeze.

YOU WILL MAKE

In this project you're going to make the micro:bit show a *sneeze* whenever the Cook comes close with her pepper pot.

The trick is that the Cook is holding a pepper box containing a magnet. The micro:bit can detect when a magnet comes close with its magnetometer sensor.

YOU WILL LEARN

This chapter introduces working with the micro:bit **magnetometer sensor** which can detect magnets and some metals. It introduces making with **Origami**.

YOU WILL NEED

ESSENTIALS

- ♣ micro:bit (USB cable and battery pack)
- ♣ Paper (2 sheets)
- ♣ Magnet (experiment with different ones)
- ♣ Scissors, glue, marker pen, ruler

GET CREATIVE

- ♣ Origami paper (2 sheets)

TEMPLATES

🖨 **Print** a template from the website on paper to make an Origami box and lid.

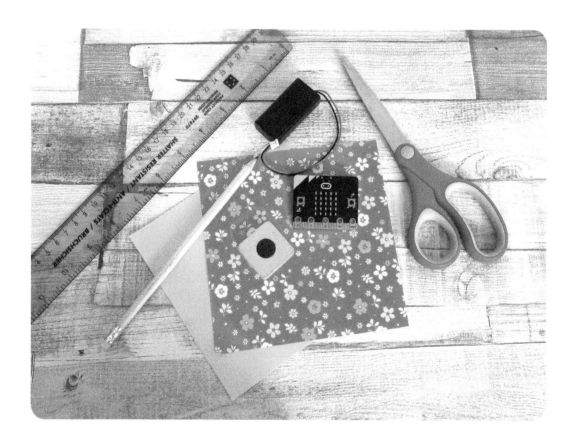

CRAFT

MAKE A PEPPER BOX

Instructions on how to fold the Origami box is available on the template.

🌐 **Find** a video to help you fold the box on the website.

Alternatively you could use **Origami** paper or make two square sheets of paper. One sheet needs to be slightly smaller. Cut 1.5cm off two adjacent sides on one of the sheets. You're going to fold two boxes, one slightly smaller and the larger box becomes the lid.

> **Origami** is the Japanese art of folding paper into decorative objects.

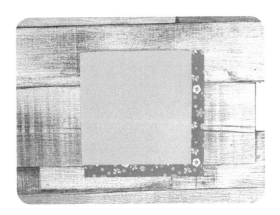

Put a small magnet inside your box and use the second, smaller box to secure the magnet inside. Any fridge magnet should work. Experiment with different magnets.

You can make a label with PEPPER written on it to decorate the box lid.

⚠ **Warning!** Don't put a strong magnet near the computer you are using to program the micro:bit.

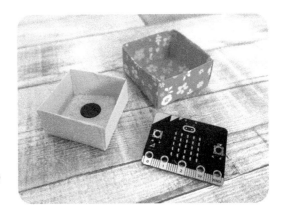

CODING

READ THE MAGNETOMETER

The micro:bit has a built-in **magnetometer** which can measure a magnetic field.

> A **magnetometer** measures the strength
> and direction of a magnetic field.

First let's print out the value reported from the magnetometer and see how it behaves when a magnet comes close.

> *Add* the magnetic force block by clicking
> ...More in the Input section.

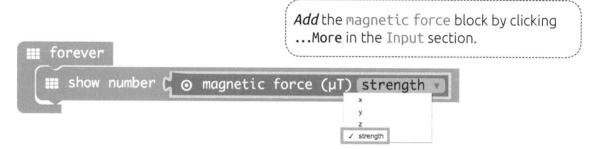

⊟ **Note:** You have to choose **strength** which is measured in microteslas — the unit of strength of magnetic fields.

⬇ **Download** the program to your micro:bit.

USING THE MAGNETOMETER

When you're using the magnetometer you need to calibrate it first to make sure the readings are accurate.

The micro:bit will tell you to DRAW A CIRCLE and then one LED will flash.

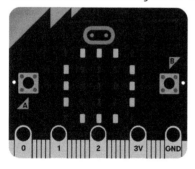

> Tilt the micro:bit around until the LEDs are lit up in a circle.
> It takes some practice!
> The micro:bit will show a smiley face when calibration is complete.

⊟ **Note:** You need to calibrate the magnetometer every time you download new code to the micro:bit.

⚠ **Warning:** Make sure you keep the micro:bit away from magnets and metal objects when you calibrate it.

TEST THE READINGS

Look at the number shown on your micro:bit. It will probably change a little each time it's shown.

Now hold a magnet and move it close to the micro:bit. Notice how the number changes. When a magnet is close the number becomes bigger.

DETECT WHEN A MAGNET IS CLOSE

Let's make the micro:bit say *Ah Choo!* when a magnet is close. You'll use 100 microteslas as the threshold.

Change your code as shown below:

```
forever
    if    magnetic force (µT) strength  > ▾  100
    then  show string  " Ah Choo! "
```

⬇ **Download** the code onto the micro:bit and remember to calibrate it.

TEST THE SNEEZE

Bring the pepper box, with your magnet inside, close to the micro:bit and check that you see *Ah Choo!* on the display.

You can adjust the value **100** and see what works best for your magnet. The sneeze can also be triggered by some metal objects. So radiators, table frames and even computers may interfere with the pepper box detection.

⚠ **Warning:** Calibrate and test your pepper box away from metal objects.

ANIMATE THE SNEEZE

Instead of text you can animate the sneeze. The micro:bit's display is made up of LEDs— really tiny ones! You can program them to create an animation.

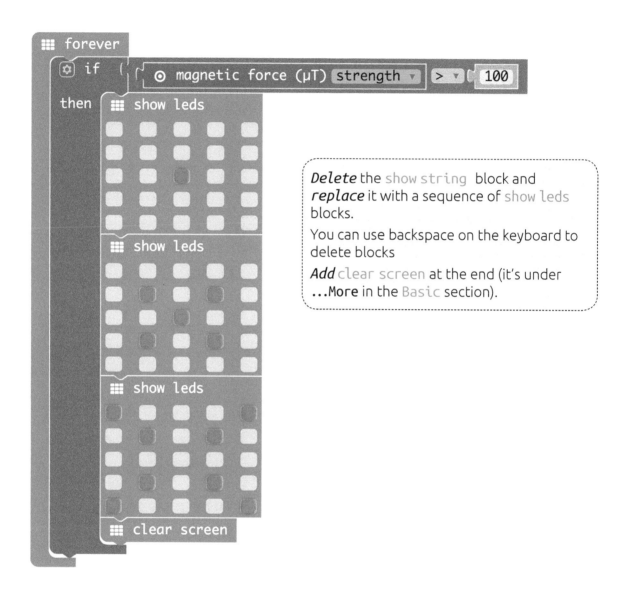

Delete the show string block and *replace* it with a sequence of show leds blocks.

You can use backspace on the keyboard to delete blocks

Add clear screen at the end (it's under **...More** in the Basic section).

⬇️ **Download** the program to your micro:bit and test your animation by bringing the pepper box close to the micro:bit.

A BIT OF HISTORY

ABOUT CARL FRIEDRICH GAUSS

The first magnetometer was invented by Carl Friedrich Gauss in 1833. In some countries magnetic fields are measured in Gauss.

ABOUT NIKOLA TESLA

The microtesla (µT) is named in honour of Nikola Tesla who made major contributions to our understanding of electromagnetism. Tesla was born in 1856 and would have been 9 years old when *Alice's Adventures in Wonderland* was published.

MAKE IT YOURS

Share your make #techalice

Customise or improve on your project:

- ♣ Improve your Origami skills by making another box with patterned paper. Perhaps, try a different Origami box design?

- ♣ Create characters from playing cards or make a figure from card and attach the micro:bit as the head so that the character sneezes when the pepper box comes close.

- ♣ Add an image to show when the character is not sneezing and change the sneeze animation.

'"I didn't know that Cheshire cats always grinned; in fact, I didn't know that cats could grin."'

THE CHESHIRE CAT

CHAPTER 6

FOLLOW THE STORY

📖 **Read** Chapter 6 *Pig and Pepper* where you meet the Cheshire Cat for the first time.

The Cheshire Cat in *Alice's Adventures in Wonderland* has a huge grin. The Cheshire Cat often disappears, leaving only its grin. *"'All right," said the Cat; and this time it vanished quite slowly, beginning with the end of the tail, and ending with the grin, which remained some time after the rest of it had gone.'*

YOU WILL MAKE

In this project, you will draw a picture of a cat on the micro:bit and then gradually remove pixels, starting from its tail, so that only a grin is left.

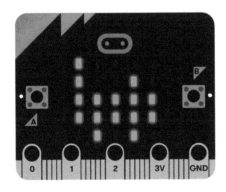
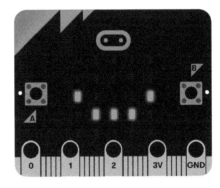

YOU WILL LEARN

This chapter introduces **plotting** and **unplotting** individual **pixels** on the micro:bit display using **coordinates**. It also focuses on **sequencing** (ordering of instructions) and working with time delays using **pause**.

YOU WILL NEED

♣ micro:bit (USB cable)

CODING

DRAW A CAT

The micro:bit has 25 **pixels**. Computer screens usually have a lot more pixels!

> A **pixel** is a single dot on a computer screen.

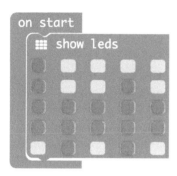

> *Create* this image of a cat using a show leds block. You'll have to use your imagination!

REMOVE THE CAT'S TAIL

The pixels on the micro:bit are numbered from (0, 0) which is the top-left (the top of the cat's tail) to (4, 4) which is bottom-right.

The first number is the x **coordinate** which runs along the horizontal axis (left to right) of the display.

The second number is the y **coordinate** which runs up along the vertical axis (up and down) of the display.

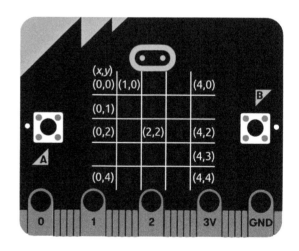

> A **coordinate** is a set of values (x,y) that show an exact position on a map, graph or display.

You can create a delay before the next instruction is executed by using the pause block. The micro:bit measures a pause in **milliseconds** (ms).

> There are 1000 **milliseconds** (ms) in one second.

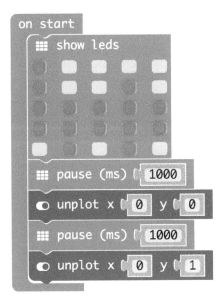

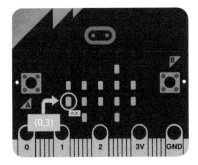

Add an unplot block (in the Led section) to remove the pixel at the top of the cat's tail after one second.

Then the next pixel of the tail after another one second delay.

🚏 **Remember:** You can duplicate blocks to save time.

JUST LEAVE A SMILE

Can you remove more pixels from the cat so that you are left with just a smile? Add more code blocks to remove all the LEDs except a smile.

Hover over a pixel in the simulator to find its coordinates. You want to be left with LEDs (2, 0); (3, 1); (3, 2); (3, 3) and (2, 4) still on.

🚏 **Hint:** The next pixel to remove is (0, 3).

⬇️ **Download** your animation to the micro:bit to see it play.

💬 *"'Well! I've often seen a cat without a grin," thought Alice; "but a grin without a cat! It's the most curious thing I ever saw in my life!'"*

Share your make #techalice

MAKE IT YOURS

♣ Can you improve the cat animation and be left with a smile?

♣ Can you make the cat slowly reappear?

♣ Try changing the timing. You could add a variable to change the delay.

♣ Can you program an input, such as pressing button A, to start the animation or make the cat reappear?

'There was a table set out under a tree in front of the house... "There's plenty of room!" said Alice indignantly, and sat down in a large arm-chair at one end of the table.'

TWINKLE TWINKLE LITTLE BAT

CHAPTER 7

FOLLOW THE STORY

📖 **Read** Chapter 7 *A Mad Tea-Party* and learn about the most peculiar tea party.

Alice meets the March Hare, the Hatter and a Dormouse at a tea party set under a tree. At the Mad Hatter's tea party the Hatter recites a rhyme:

"Twinkle, twinkle, little bat!
How I wonder what you're at!
Up above the world you fly,
Like a tea tray in the sky.
Twinkle, twinkle, little bat!
How I wonder what you're at!"

YOU WILL MAKE

In this project you will program the buttons to play the music of *Twinkle Twinkle Little Bat* on the micro:bit and connect some headphones or a buzzer to hear it play. You'll press the buttons—**A B B A**—to play the whole tune.

YOU WILL LEARN

This chapter introduces programming with **musical notes** and connecting up headphones, a buzzer or small speaker as an output on the micro:bit. It also uses the A and B **button** inputs on the micro:bit. Reading of **musical notation** is also explained and used.

YOU WILL NEED

ESSENTIALS

- ♣ micro:bit (USB cable and battery pack)
- ♣ Headphones
- ♣ 2 crocodile clip leads

GET CREATIVE

- ♣ Card (black)
- ♣ Buzzer (one that can play different tones) or small speaker
- ♣ MI:power board (optional)
- ♣ Scissors, double-sided tape, sticky tape
- ♣ Elastic band (or a chain of loom bands)
- ♣ Decorative material (ribbon, coloured paper, feathers, etc.)

TEMPLATE

🖶 **Print** the mini top hat template from the website on paper/card. You can trace the template onto black card to make a more authentic looking top hat.

🌐 **Find** a template to cut the top hat shape with a craft cutter on the website.

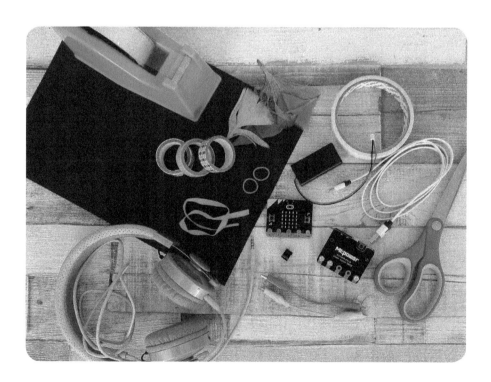

PROGRAMMING MUSIC

The micro:bit can play musical notes. You can choose the note that is played and how long it's played for (the beat). Before you start programming with music you need to understand some musical notation. If you can already read music you can apply your knowledge.

MUSICAL NOTATION

The micro:bit has 3 octaves to program—the low, middle and high octaves. Each octave has 7 notes—C, D, E, F, G, A, B.

The middle note on the keyboard is called `Middle C`. You'll be programming the middle octave and select notes using a keyboard.

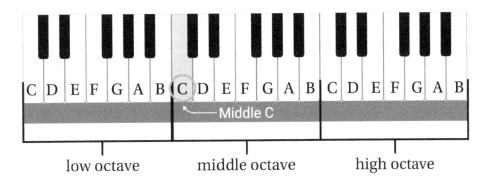

Music notation shows notes on staves (horizontal lines) and its position tells you the name of the note, eg. C.

A bar is a specific number of beats and separated with a vertical line. In *Twinkle, Twinkle Little Star* each bar has 4 beats.

Notes can have different durations. In the below example, filled in notes are 1 beat long (crotchet/quarter note) and the note with a hole is two beats long (minim/half note).

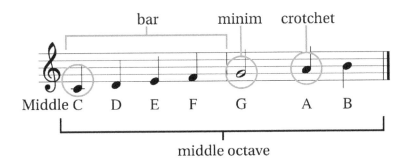

CODE *TWINKLE TWINKLE LITTLE BAT*

You're going to program the input buttons **A** and **B** to play the tune. Start by programming the first 2 bars of the tune.

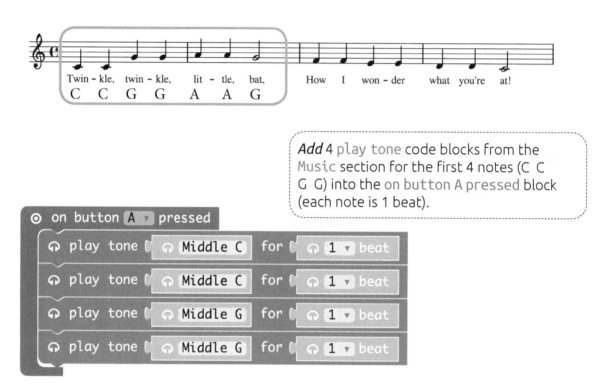

> *Add* 4 play tone code blocks from the Music section for the first 4 notes (C C G G) into the on button A pressed block (each note is 1 beat).

Duplicate the play tone block to add the next 3 notes (A A G) from the second bar. Did you spot the last G is 2 beats? You should now have 7 play tone blocks all together.

▶ **Test** the tune by clicking button **A** in the simulator. It uses your computer to play the sound.

Before you continue to program the rest of the tune, you're going to try play the sound through an external output such as headphones or a buzzer.

⤓ **Download** your program and transfer onto the micro:bit.

CONNECTING SOUND

The micro:bit doesn't have a speaker but it can generate sounds if you connect headphones, a buzzer or small speaker to the micro:bit.

CONNECTING HEADPHONES

⚠ **Notice:** Connect two crocodile clip leads to the micro:bit—one to Pin 0 and the other to GND. Then connect the other ends to the right parts of the headphone jack as shown below.

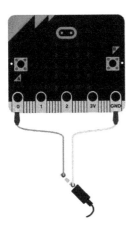 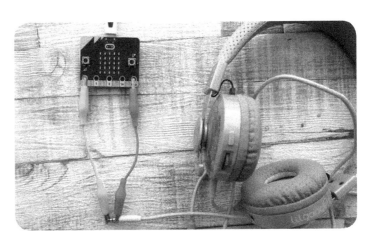

If your micro:bit is still plugged into the computer, mute the sound so you can hear the sound through the headphones. Alternatively you can use the battery pack to power the micro:bit and play the sound.

⚠ **Safety:** You can't control the volume on the micro:bit so use volume-control headphones. You should be able to hear the music without actually wearing the headphones.

CONNECTING A BUZZER

Alternatively, you can use a buzzer or simple speaker.

If your buzzer is marked with positive and negative then connect the GND crocodile clip to the negative leg or wire and the Pin 0 crocodile clip to the positive leg or wire, otherwise it should work either way.

Depending on the buzzer, you might have to listen carefully to hear the sound.

 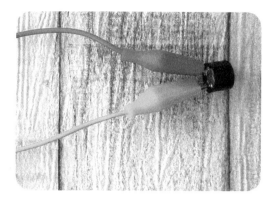

It's good practice to test your code before programming too many notes. Press button **A** and you should hear the micro:bit playing music through the headphones, buzzer or speaker.

🐞 **Troubleshoot:** If you don't hear anything try putting a show icon (music note) before the first note and a clear screen block after the last note so that you know when the music should be playing.

CODE *HOW I WONDER...*

You've already programmed the first two bars of the rhyme. Now add more play tone blocks in the on button A pressed block to program the next 2 bars of the rhyme—*'How I wonder what you're at!'*.

🗃 **Hint:** You'll need 14 play tone code blocks for the music below.

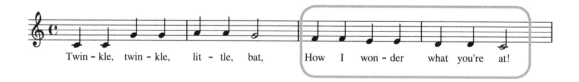

CODE *UP ABOVE...*

Use the musical notation below to program the next 2 bars of the rhyme—*'Up above the world you fly,'* in an on button B pressed block.

The code blocks below will get you started!

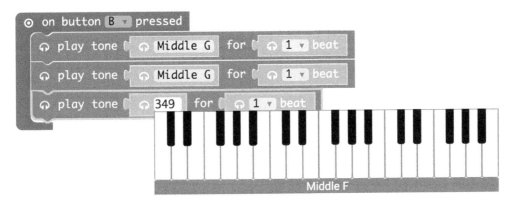

🎼 **Think:** Did you spot the first 2 bars have the same notes as the second 2 bars? You can just press B again to repeat the tune to play—*'Like a tea tray in the sky'*.

CODE THE REST OF THE RHYME

The last two lines of the rhyme repeat the first two lines both in tune and words. If you press button A it should play the following 4 bars of music:

▶ **Try** playing the whole tune by pressing buttons **A** and **B** in the simulator in the correct order (**A B B A**).

PLAY THE TUNE

⬇ **Download** your program to the micro:bit, disconnect it from your computer and attach the battery pack.

Connect the micro:bit to your chosen sound output and play the whole tune by pressing the A and B buttons (**A B B A**).

MAD HATTER'S WEARABLE

Make a papercraft top hat to house the micro:bit and play the tune. Instructions are available on the template.

Attach the buzzer, as previously described, using two crocodile clip leads. Secure the micro:bit to the top hat with loom bands and conceal the battery pack inside the hat.

🔖 **Tip:** You can also use the MI:power board to power the micro:bit and play the sound through its built-in buzzer.

A BIT OF HISTORY

ADA LOVELACE PROGRAMMING MUSIC

Ada Lovelace was the first person to realise the power of computer programming. There were no computers in her time, her ideas were based on Charles Babbage's Analytical Engine which he didn't manage to build. She predicted that computers would be able to compose music.

"Supposing, for instance, that the fundamental relations of pitched sounds in the science of harmony and of musical composition were susceptible of such expression and adaptations, the engine might compose elaborate and scientific pieces of music of any degree of complexity or extent." Countess Ada of Lovelace, 1843.

"At any rate I'll never go there again!" said Alice as she picked her way through the wood. "It's the stupidest tea-party I ever was at in all my life!"

MAKE IT YOURS

Share your make #techalice

♣ Can you add images or animations to the micro:bit to go with the tune?

♣ Can you use other inputs to start playing the tune without having to press the buttons?

♣ Can you make the micro:bit also scroll **10/6**—the Mad Hatter has a price label on his hat as he borrowed it from his shop.

> **10/6** (say 10 and 6) means ten shillings and sixpence in pre-decimal British currency.

♣ Make the top hat using felt and practice your sewing skills.

'A large rose-tree stood near the entrance
of the garden: the roses growing on it were
white, but there were three gardeners at it,
busily painting them red.'

PAINTING THE ROSES

CHAPTER 8

FOLLOW THE STORY

📖 **Read** Chapter 8 *The Queen's Croquet-Ground* where you meet three gardeners painting white roses red, right at the beginning of the chapter.

The gardeners (spades) paint the white roses red to avoid the anger of the Queen of Hearts. When the Queen comes along they all lie face down on the ground so that the Queen can't tell who they are—all cards look the same from the back.

💬 *'Why the fact is, you see, Miss, this here ought to have been a red rose-tree, and we put a white one in by mistake; and if the Queen was to find it out, we should all have our heads cut off, you know.'*

YOU WILL MAKE

In this project, you'll code a game where you need to react to what appears on the micro:bit screen. If it's a rose, **shake** the micro:bit to paint it, if it's a heart *flip* the micro:bit face down to hide from the Queen! See how high a score you can get by painting as many roses as you can without the Queen spotting you.

YOU WILL LEARN

This chapter introduces working with **random** numbers and working with the **on screen up/down** events on the micro:bit. There's also the opportunity for crafting and looking at the **design** and **usability** of a game. **3D printing** can also be used.

YOU WILL NEED

ESSENTIALS

- ♣ micro:bit (USB cable and battery pack)
- ♣ 2 loom bands
- ♣ 2 playing cards (use two, five or seven of spades)
- ♣ Sticky tape

GET CREATIVE

- ♣ Black air drying clay
- ♣ Black card
- ♣ 3D printer (black filament to 3D print a head and feet)

TEMPLATE

⊕ **Find** a template of a spade head for a craft cutter on the website. (You can also cut your design with scissors from black card.)

You'll be coding the game in stages so complete each step fully before moving on to the next part of the game code. This will help you understand what is happening.

HIDE FROM THE QUEEN!

You're going to be looking at the micro:bit screen a lot in this game so let's turn the brightness down.

The first part of the game is to hide from the Queen. When a heart appears on the screen you need to turn the micro:bit screen down to hide from her. The screen should be blank when you turn it back over. You're going to use a **while loop**.

> A **while loop** repeats the code inside it while a condition is true.

At the start of each repetition the condition is checked, if the condition evaluates to true then the code inside the loop will repeat, if the condition is false then the loop finishes and any code after the loop will run.

First let's make the Queen to keep appearing forever. You can use a **while true loop**.

> A **while true loop** will repeat forever because the condition is always true.

Add this code. Change the value to a smaller number on the set brightness block (it's under ...**More**) to reduce the brightness of the LEDs. To display LEDs use the show icon block. You'll be using the **heart** icon.

Add a clear screen block in an on screen down event block to make the heart icon disappear when you turn the micro:bit on its face.

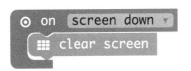

⬇️ **Download** and transfer your program to the micro:bit to test your code.

TEXTING *OWHH*

OWHH is what the Queen of Hearts would text if she had a mobile phone! It's short for *'off with his head'* or *'off with her head'*.

Now you're going to add code so that the player has to react when a heart appears on the screen meaning that the Queen has arrived.

```
on start
    set brightness ( 15
    set queen to ( false
    while ( not ( queen
    do     set queen to ( true
           show icon [heart]
           pause (ms) ( 2000

    clear screen
    show string ( " OWHH "
```

> **Add** this code to your program. Create a new Boolean variable queen.
>
> Setting the queen variable to true and displaying a heart on the micro:bit means that the Queen is coming and a 2-second timer is started.
>
> Turning the micro:bit over (screen down) sets queen to false. If queen is still true when the timer finishes, the loop won't repeat and the code immediately after it will run, so you'll see the string—OWHH.

```
on screen down
    set queen to ( false
    clear screen
```

∴ **Think** about the code and make sure you understand it.

⬇ **Download** and transfer your program to the micro:bit. Test it. Practice how far you have to turn the micro:bit to remove the Queen as it will become harder later.

RANDOM PICTURE

Now you need to add some roses to paint. The player needs to shake the micro:bit to paint each rose that appears. You want a rose to appear more often than the heart.

Pick a **random number** between 0 and 4. If it's 0 then show a heart, otherwise (if it's 1, 2, 3 or 4) show a rose. This means that one turn in 5 you'll see a heart.

Random numbers don't follow a pattern. They are useful for programming unpredictable behaviours.

Change your code in the on start block by adding an if..then..else logic block in the **do** part of the while do loop block.

Create a new variable called rose and draw a rose in a show leds block.

Note that you need to move the pause block to after the if..then..else block.

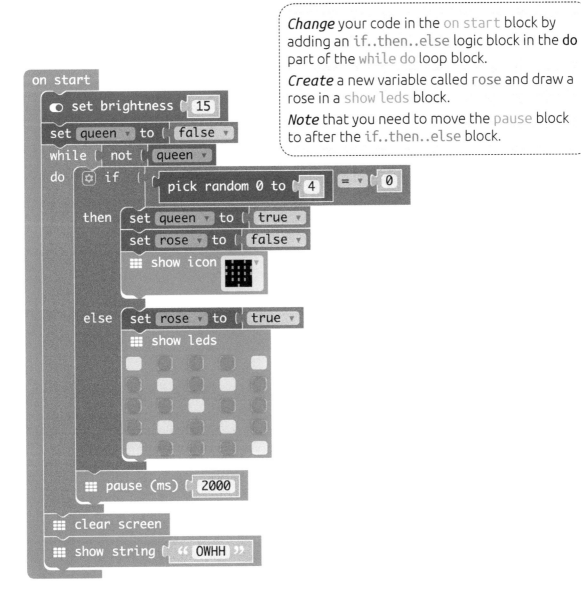

Check carefully that your code in the on start block looks as the above code. You'll be adding more code to this block later.

```
⊙ on  screen down ▾
    ⚙ if  ( queen ▾
    then  set queen ▾ to ( false ▾
          ⊞ clear screen
```

```
⊙ on  shake ▾
    ⚙ if  ( rose ▾
    then  set rose ▾ to ( false ▾
          ⊞ clear screen
```

🏳 **Hint:** Did you spot you also had to add an if..then logic block to your on screen down event?

⤓ **Download** the program to your micro:bit and test it.

A BIT OF HISTORY

TELEGRAPH - TEXT MESSAGES

There were no text messages or instant chat apps in Victorian times. You could send a short text message called a telegram to someone using the Telegraph Service. Sir William Fothergill Cooke and Charles Wheatstone pioneered telegraphy in Britain while Samuel Morse worked on similar ideas in the US. Communication between the countries took a long time in those days! Around the time *Alice's Adventures in Wonderland* was published, a successful telegraph cable was laid under the Atlantic from Ireland to Newfoundland and messages between Europe and the Americas took hours rather than weeks by ship.

ADD A SCORE

Now you can add a score variable to keep track of how many roses you manage to paint before the Queen spots you.

Add set score to 0 in your on start block. This will store the number of roses you have successfully painted.

...

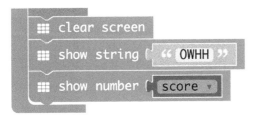

Add a show number block to display the score of the game.

Add code to increase the score by 1 whenever you *paint a rose*.

⤓ **Download** and transfer your program to your micro:bit and test your game.

How many roses can you paint before the Queen spots you?

MAKE A GARDENER

In *Alice's Adventures in Wonderland* the spades are the gardeners and are responsible for painting the roses.

You'll need either the 2, 5 or 7 spade card and any other card to turn your micro:bit into a gardener. Don't use the club playing cards as you'll need them in another project.

Use the loom bands to attach the micro:bit to the front of a spade card. Connect the micro:bit to a battery pack and put it behind the bottom of the card. Place another card on top of the battery pack with its front facing inwards.

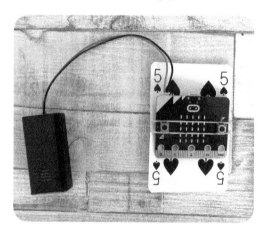

Use some sticky tape to keep the cards and the battery pack together to create a card that can stand up or lie down.

Use scissors to cut out a spade-shaped head from black card. Add it to the gardener using sticky tape.

You could also make a head and feet using black air drying clay or try 3D printing the playing card's head and feet.

⊕ **Find** details on the website to 3D print a head and feet.

💬 *'... "The Queen! The Queen!" and the three gardeners instantly threw themselves flat upon their faces. There was a sound of many footsteps, and Alice looked round, eager to see the Queen.'*

Share your make #techalice

MAKE IT YOURS

Customise your game and decide how you want it to work.

♣ If the game is too hard or too easy you can adjust the amount of time that you have to lie down when you spot the Queen.

♣ Do you want the game to work differently? Change it!

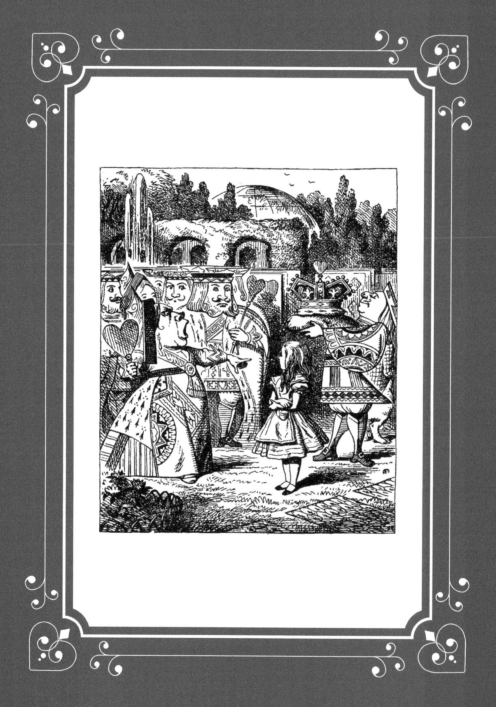

'The Queen had only one way of settling all difficulties, great or small. "Off with his head!" she said without even looking round.'

PUSHING THE QUEEN'S BUTTONS

CHAPTER 9

FOLLOW THE STORY

📖 **Read** Chapter 8 *The Queen's Croquet-Ground*, where you meet the angry Queen of Hearts. If you've already read this chapter then there's no further reading for this project.

T he Queen of Hearts becomes angry rather easily in the story. To *push someone's buttons* means to make them angry.

💬 *'... in a very short time the Queen was in a furious passion, and went stamping about, and shouting, "Off with his head!" or "Off with her head!" about once in a minute.'*

YOU WILL MAKE

In this project, you will create an Anger Monitor to warn the inhabitants of Wonderland about the Queen's current anger level.

The micro:bit has two buttons labelled **A** and **B** which you will use to set the Queen's anger level by lighting up LEDs.

You can make the Anger Monitor into a paper or e-textiles tabard for a card character.

YOU WILL LEARN

This chapter introduces working with **multiple LED outputs** and developing a **monitor** to quickly show information. It involves creating a good **algorithm** to solve a problem. The project includes crafting and optionally sewing **e-textiles** (working with sewable electronics) and it also introduces **concurrency**.

YOU WILL NEED

ESSENTIALS

- ♣ micro:bit (USB cable and battery pack)
- ♣ 4 crocodile clip leads (useful to have one red, amber and green lead but not essential)
- ♣ 3 LEDs (standard or sewable LEDs)

GET CREATIVE

- ♣ 2 playing cards
- ♣ 2 loom bands
- ♣ Conductive thread, felt, embroidery thread (to make an e-textiles tabard)
- ♣ Conductive tape, paper/card (to make a paper tabard)
- ♣ 4 crocodile clips (for the paper/e-textiles tabard)
- ♣ 10 mini paper fasteners (for the paper tabard)
- ♣ Scissors, sticky tape, pliers
- ♣ Black card (or use the head and feet you made in Chapter 8)

TEMPLATES

🖶 **Print** a template from the website for the paper tabard or a pattern to make an e-textiles tabard.

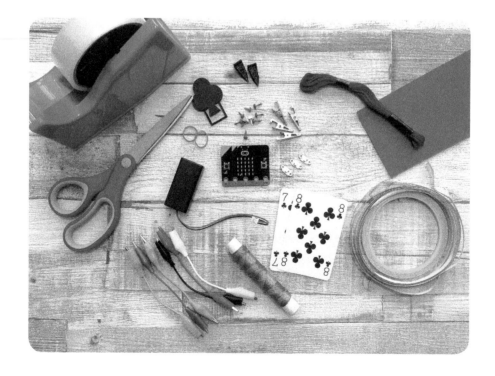

CHOOSING THE RIGHT LEDS

You will need 3 LEDs for this project.

LED OPTIONS

You could use 3 LEDs of the same colour and light up more of them as the Queen becomes angrier. Purple is a good choice as we also say someone goes *purple with rage*. Purple is also a royal colour and makes a change from the red LEDs on the micro:bit.

Alternatively, you could use 3 different colours such as green, amber and red.

SAFETY WARNING

Only use LEDs that you know are suitable for the micro:bit. LEDs normally require a resistor so that they don't have too much electrical current pass through them. This could damage the LED or even make it explode. If your LEDs came in a micro:bit kit the instructions should tell you which resistor, if any, to use.

LEDs with built-in resistors designed for use in sewable projects with coin cell batteries or 3V microprocessors should be suitable for the micro:bit.

OUR RECOMMENDATION

In this project we used standard LEDs from an e-textiles kit, which is compatible with 3V boards and doesn't require a resistor. We also used sewable LEDs which have built-in resistors.

⊕ **Find** more information on the website about e-textile kits and sewable LEDs.

LEVELS OF ANGER

You are going to program button **A** to show when the Queen becomes angrier and button **B** when she calms down.

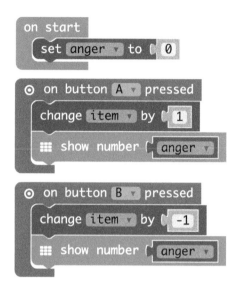

> *Add* this code including a variable called anger to show how angry the Queen currently is.

⊙ **Test** your code in the simulator.

MORE ANGRY

Now you are going to use LEDs to clearly show how angry the Queen is. You need an **algorithm** to do this.

> An **algorithm** is a sequence of precise rules or instructions that does useful things.

When button **A** is pressed to increase the Queen's anger level, you need to light up the next LED. In this case, the problem is how to light up the correct LEDs when a button is pressed. First, let's look at what needs to happen:

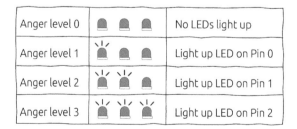

Anger level 0	🔔 🔔 🔔	No LEDs light up
Anger level 1	🔔 🔔 🔔	Light up LED on Pin 0
Anger level 2	🔔 🔔 🔔	Light up LED on Pin 1
Anger level 3	🔔 🔔 🔔	Light up LED on Pin 2

This algorithm means that you don't need to change the LEDs that are already lit up.

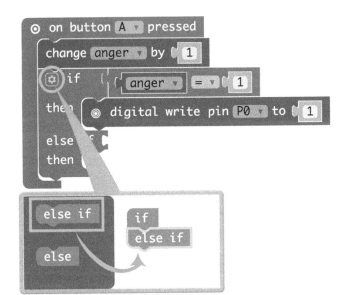

You need to add several `if..then` logic blocks to light up the LEDs.

Add more `else..if` conditions by clicking the **gear** icon.

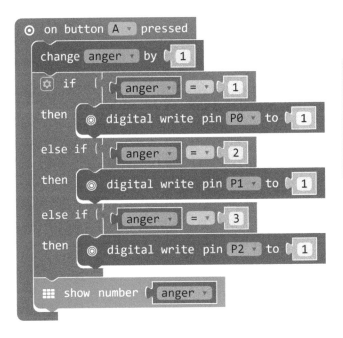

Add this code to your on button A pressed block to turn pins *on* (1) as the Queen becomes more angry.

Remember the Pin blocks are in the Advanced section.

▶ **Test** this code in the simulator by pressing the **A** button to see the pins light up. The angrier the Queen becomes the more output pins light up.

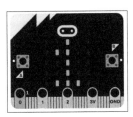   

∴ **Think** about how this code will work and make sure you understand it.

LESS ANGRY

The Queen might become less angry. She soon forgets about things.

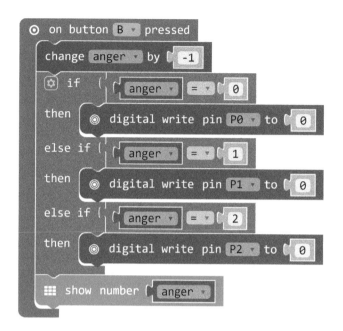

> *Add* this code to on button B pressed to make the Queen less angry.
>
> This time you'll be turning pins *off* (0) rather than *on*.

If you press button **B** when the Queen's anger level is 2 then it will go down to 1 and Pin 1 which controls the second LED will be turned off.

▶ **Test** your code in the simulator and make sure you understand how pressing the buttons turns the outputs *on* and *off*.

⬇ **Download** your code ready to connect the LEDs. Disconnect your micro:bit from the computer and attach the battery pack.

CONNECTING LEDS

Now it's time to connect up your LEDs. We're going to show you two different methods.

USING STANDARD LEDS

You'll need 3 standard LEDs and 4 crocodile clip leads.

⚠ **Warning:** Check whether your LEDs need resistors.

You need to connect the short leg of each of your LEDs to ground (GND on the micro:bit). Twist the short legs of your LEDs together (it's easier using pliers) and connect them all with a single crocodile clip lead to GND.

Connect the long legs of the LEDs to Pin 0, Pin 1 and Pin 2 as shown below.

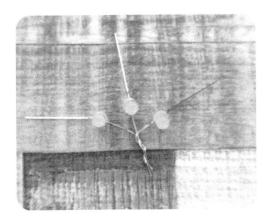 

⚠ **Warning:** Be careful that the LED legs only touch where they need to.

USING SEWABLE LEDS

You can also use sewable LEDs. Connect the negative sides of the LEDs to GND and the positive sides to Pin 0, Pin 1 and Pin 2 using crocodile clip leads as shown below.

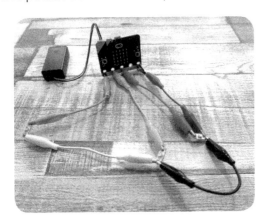 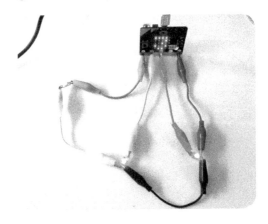

Later you can use the sewable LEDs to make a wearable for a playing card character.

TRY IT OUT

Press buttons **A** and **B** on the micro:bit. The LEDs should light up in sequence as the Queen becomes angrier and turn off as she becomes less angry.

OFF WITH HIS/HER HEAD

You know what happens if the Queen's anger level goes over 3!

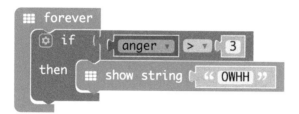

When the anger variable is over 3 the micro:bit will scroll —OWHH— (text speak for *off with her/his head*).

Check that the anger variable is greater than three rather than equal to 4 so that the Queen remains angry if you make anger go higher than 4.

Add an *additional* forever block so that the scrolling text and flashing lights happen at the same time.

This code will make the LEDs flash! Use the forever block so that it repeats if the anger level stays above 3.

You should now have two forever blocks running at the same time. This is called **concurrency**.

> When things happen at the same time in computing we call it **concurrency**.

You can make a tabard for a playing card using paper or fabric. We'll show you how to do both! You'll be using sewable LEDs for both methods.

PAPER TABARD

You'll need paper/card, conductive tape/kitchen foil, 4 crocodile clips, 10 mini paper fasteners and 3 sewable LEDs to make the paper tabard.

Cut out the paper tabard from the template. Stick conductive tape as shown on the template.

Attach your sewable LEDs using mini paper fasteners by pushing them through the card and sewable LED.

Make sure you have the positive and negative sides the right way round and make firm connections.

Now connect 4 crocodiles clips also using the mini paper fasteners as shown below. Pliers are useful to flatten the metal prongs on the clips.

Tip: Make sure the mini clips don't touch at the back. Use some sticky tape to prevent them from spinning round and accidentally touching. The coating on the paper fasteners isn't conductive but the metal underneath is.

E-TEXTILES TABARD

You'll need a piece of felt, conductive thread, 4 crocodile clips and 3 sewable LEDs to make the e-textiles tabard.

Trace the pattern from the template onto your felt. Refer to the circuit diagram on the template and then use conductive thread to sew the crocodile clips and sewable LEDs to the felt.

Use regular non-conductive embroidery thread to add decorative details to the tabard.

Attach the micro:bit by clipping the crocodile clips to the correct pins. Attach the battery pack and dress a club playing card character with your tabard. (Make a standing card character like you did in Chapter 8.) Use a loom band to keep the micro:bit secure as shown below.

'The Queen turned crimson with fury, and, after glaring at her for a moment like a wild beast, screamed "Off with her head! Off—" '

MAKE IT YOURS

Share your make #techalice

Customise the anger meter so it works the way you want it to.

♣ Maybe you want to only light up one LED at a time so that the first LED goes off when the second one comes on?

♣ Maybe you want to show more levels of anger? For example, you might just use the first LED for levels 1-3 but make it flash faster as the Queen becomes angrier.

♣ Can you design and make a wearable for yourself to wear instead of the card? You could use it to warn others what kind of mood you're in!

♣ Would you prefer it if the **A** button decreased the Queen's anger and the **B** button increased it? Change it.

♣ Can you stop the anger level going negative and maybe add a maximum level of anger?

"'I don't think they play at all fairly," Alice
began, in rather a complaining tone, "and
they all quarrel so dreadfully one can't
hear oneself speak —"'

THE CROQUET MATCH

CHAPTER 10

FOLLOW THE STORY

📋 **Read** Chapter 8 *The Queen's Croquet-Ground* to learn about the strange croquet game Alice plays. If you've already read this chapter, get started with the project!

I n *Alice's Adventures in Wonderland*, Alice joins the Queen and King and other players in a croquet match of sorts.

💬 *'Alice thought she had never seen such a curious croquet-ground in her life: it was all ridges and furrows: the croquet balls were live hedgehogs, and the mallets live flamingoes, and the soldiers had to double themselves up and stand on their hands and feet, to make the hoops.'*

YOU WILL MAKE

In this project, you will create a croquet scoreboard. There are different versions of croquet but you'll use Golf Croquet which is simpler. You will also make the hedgehog balls, flamingo mallets and playing card hoops so that you can actually play the game.

YOU WILL LEARN

This chapter introduces the **tilt left** and **tilt right** micro:bit input events. The coding in this project is simple so you can concentrate on getting creative with the craft!

YOU WILL NEED

ESSENTIALS

- ♣ micro:bit (USB cable and battery pack)
- ♣ 6 playing cards (club cards A-6 to match the hoops)
- ♣ 2 playing cards (any others to make a scoreboard)
- ♣ Black card, paper
- ♣ Loom bands
- ♣ 2 wooden craft sticks
- ♣ 4 marbles
- ♣ Scissors, glue

GET CREATIVE

- ♣ Decorative materials (googly eyes, feathers, etc.)
- ♣ Pipe cleaners (pink, black, blue, yellow, red)
- ♣ Air drying clay (red, black, yellow, blue)

TEMPLATE

🖨 **Print** a template from the website for the playing card characters and flamingo mallets. There are 2 versions of the flamingo template to choose from.

🌐 **Find** a template on the website to cut the flamingo and card character shapes with a craft cutter.

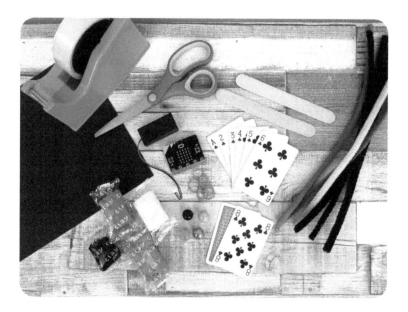

THE GAME

In Golf Croquet there are two teams and they play with 4 balls - blue, black, red and yellow. The players take turns to hit the balls with a mallet through a sequence of 6 ordered hoops in the right direction.

Each team, let's call them **Team A** and **Team B**, plays with two balls each. Team A plays a blue ball and a black ball and Team B plays a red ball and a yellow ball. If you only have two players then they alternate between their two colours on each turn.

The first player to hit their ball through a hoop scores a point for their team and all the players move on to the next hoop. The first team to reach 4 points is the winner. If both teams have 3 points after hoop 6 then hoop 2 is played again in the reverse direction as a decider. This is the short (7 point) version of the game.

A BIT OF HISTORY

HISTORY OF CROQUET

The game of Croquet became amazingly popular after it was shown at the Great Exhibition at Crystal Palace, London in 1851, not long before Lewis Carroll featured it in his book. Croquet spread throughout the country like loom bands and fidget spinners have in more recent times.

CODING

THE DISPLAY

You will program the micro:bit display to show the current hoop being played so everyone knows where they should be (unlike in Alice's game with the Queen!)

When you tilt the micro:bit to the left it will show Team A's score and when you tilt it to the right it will show Team B's score. The score is the number of hoops that have been won by that team.

When you press button **A**, Team A's score will go up by one. When you press button **B**, Team B's score will go up by one. The table shows an example game to demonstrate how it will work:

Current Hoop (Default)	Team A (Tilt left)	Team B (Tilt right)
1	0	0
2	1 (Won Hoop 1)	0
3	2 (Won Hoop 2)	0
4	2	1 (Won Hoop 3)
5	3 (Won Hoop 4)	1
6	3	2 (Won Hoop 5)
7	4 (Won Hoop 6)	2

In this example, at the end of the game the micro:bit will display *A* to show that Team A won the game. You can still tilt the micro:bit left and right to see the final team scores.

∴ **Think** carefully about how this works to make sure you understand it before you start coding the croquet game. Software engineers spend a lot of time doing **computational thinking**. Actually writing the code is just a small part of the work.

> **Computational thinking** is understanding a problem well enough that you can describe it in simple terms so that a computer can help you solve it.

KEEPING SCORE

You're going to program buttons **A** and **B** to increase the score of the corresponding team when a button is pressed.

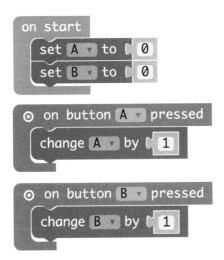

Add two variables A and B to keep track of the hoops scored by each team.

on button A pressed will add a score for team A and on button B pressed will add a score for team B.

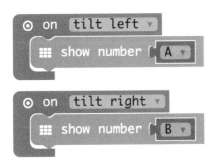

Add on tilt left and on tilt right blocks to show the scores of both teams.

🚏 **Tip:** Add a pause block to the tilt left and tilt right code to display the team's scores for longer. They disappear quite quickly.

⊙ **Test** your code in the simulator by dragging the micro:bit to the left or right.

⎙ **Download** your program to the micro:bit and try it out. To check the current scores you'll pick up the micro:bit and tilt it to the left to see Team A's score or to the right to see Team B's score.

🚏 **Practice** doing this so that you can find out how far you need to tilt the micro:bit so that a tilt left or tilt right event is generated.

HOOP CURRENTLY PLAYED

You can work out the current hoop from the players' scores. Add the scores together to know how many hoops have been played so far. Add one to show the current hoop.

> *Add* the following math blocks inside a forever block.

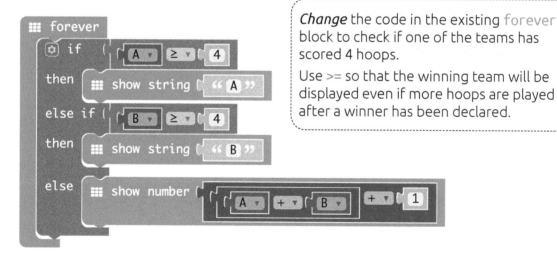

⊙ **Test** to make sure that you can still check the scores of the teams as well as viewing the current hoop.

WINNING

The winning team is the first to reach 4 points. You can display the name of the winning team on the micro:bit.

> *Change* the code in the existing forever block to check if one of the teams has scored 4 hoops.
>
> Use >= so that the winning team will be displayed even if more hoops are played after a winner has been declared.

⊙ **Test** your code fully in the simulator or on the micro:bit. Do you understand how it works?

⬇ **Download** the program to the micro:bit and attach a battery pack.

Now that we have a scoreboard, let's make the pieces for the game so you can play it.

CREATE A DISPLAY BOARD

Trace the playing card character for the scoreboard on black card and cut out.

Use a playing card to hold the scoreboard (micro:bit). You need two playing cards (don't use clubs A to 6 as you'll use them for the hoops). Stick your character to the front card and attach your micro:bit and battery pack as you did in Chapter 8.

Now you have a stand-up croquet scoreboard which should help keep track of the game. You can tilt the card to the left or right to view the scores of the teams.

MAKE HEDGEHOG BALLS

In *Alice's Adventures in Wonderland* the balls are hedgehogs! You could make balls that look like hedgehogs by rolling pipe cleaners into balls. You could just use marbles or use air-drying clay to roll your own croquet balls. These balls will all play differently—try them all and select the ones you like best for your game.

If you want to play seriously the balls need to be different colours so you know which team they belong to. The official colours for croquet are red, yellow, blue and black.

MAKE FLAMINGO MALLETS

If you printed the flamingo template on paper, stick it to card to make it sturdier.

The template has 4 flamingo shapes so that you can make two mallets. Cut out your flamingo shapes and stick them to a wooden craft stick as shown below.

Now you can be creative decorating your flamingo mallet with feathers, googly eyes, pipe cleaners (for the legs), etc.

You can also make your own flamingo mallet using air drying clay, pipe cleaners and other decorative material.

MAKE 6 HOOPS

Use the club cards numbered A to 6 to make 6 hoops. Bend the playing card over and hold it in place using one loom band.

Use the template to trace heads and feet on black card (same as the scoreboard) for your 6 characters.

Alternatively make hands and feet using black air drying clay or try 3D printing the hands and feet.

🌐 **Find** more details on the website.

The heads and feet will help the cards stand more securely.

Set up the hoops as shown below:

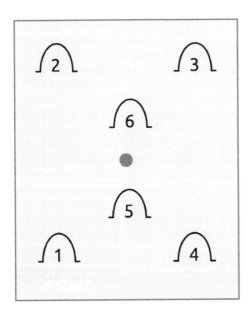

🐛 **Troubleshoot:** If you find the loom bands pinging off, use some sticky tape to keep them in place.

PLAY THE GAME

It's time to play the croquet game with the micro:bit keeping score. Ask a friend to play, or create teams with three friends taking turns to hit the hedgehog balls through the correct card hoop! Remember to keep score using the micro:bit scoreboard.

It can be tricky to play with the pipe cleaner balls (just like the hedgehogs in the book!). Agree with all players what you will do if the ball becomes stuck under a hoop.

Tip: Press the **reset** button to play another game.

"'Let's go on with the game," the Queen said to Alice; and Alice was too much frightened to say a word, but slowly followed her back to the croquet-ground.'*

MAKE IT YOURS

Ready to add your own flair? Try one or all of the challenges below.

♣ It's traditional to toss a coin to decide which team goes first. Can you program the micro:bit to decide which team goes first?

🚏 **Hint:** You'll need to use a random block.

♣ Reset the game by pressing **A+B** so you don't have to press reset when the micro:bit is attached to the cards.

♣ You can play a standard version of Golf Croquet (13 point game) where you also play all the hoops in reverse. The winner is the team that scores 7 points. If both teams have the same score after the 12th hoop, then hoop 3 is played again. Can you adapt your code for this version?

♣ Golf Croquet is a simple version of the game which hadn't been invented when *Alice's Adventures in Wonderland* was written. Can you learn the rules for Garden Croquet and invent a way to use the micro:bit to help with the game?

♣ Why not make up your own version of croquet with your choice of rules and use the micro:bit to help.

♣ How about playing solo croquet with a full-size set and attaching the micro:bit to the peg that you have to hit at the end? Start a timer on the micro:bit and then stop it when the micro:bit detects a shake from the ball hitting the peg. Try to improve on your time.

'"Thank you, it's a very interesting dance to watch," said Alice, feeling very glad it was over at last.'

THE LOBSTER QUADRILLE

CHAPTER 11

FOLLOW THE STORY

📖 **Read** Chapter 10 *The Lobster Quadrille* where the dance is described. You meet the Mock Turtle at the end of Chapter 9 *The Mock Turtle's Story*.

The Quadrille is a dance that was popular when *Alice's Adventures in Wonderland* was written. The Lobster Quadrille is based on the dance, but different! The Mock Turtle and the Gryphon explain the dance to Alice. It involves sea creatures including turtles and seals taking lobsters as partners and completing some intricate dance moves.

YOU WILL MAKE

In this project, you're going to program sprites on the micro:bit to perform a Lobster Quadrille. A precise dance or choreography is a kind of algorithm, it has a clear sequence of steps. The algorithm for a full quadrille dance is quite complicated!

YOU WILL LEARN

This chapter introduces the use of **sprites** (single pixel characters that can move around the micro:bit screen) and more advanced use of **repetition** and designing **algorithms**.

YOU WILL NEED

- ♣ micro:bit (USB cable and battery pack)
- ♣ Headphones (to program music—not required)
- ♣ 2 crocodile clip leads (to connect sound)
- ♣ Friends/teddies (to try the dance in real life)

A BIT OF HISTORY

We can trace the story of programmable **turtles** back to Alice.

> A **turtle** is also a robot or on-screen character
> that can be programmed to move around.

ABOUT TURTLES

The first robot that could move around autonomously (making decisions based on sensor input) was created by Dr William Grey Walter in the 1940s in Bristol, England. He called it a tortoise because of this scene in *Alice's Adventures in Wonderland* where the Mock Turtle says *"The master was an old Turtle—we used to call him Tortoise— ... We called him Tortoise because he taught us."*

In the 1960s Seymour Papert, Cynthia Solomon and Wally Feurzeig invented a programming language for children called Logo. Papert was inspired by Grey Walter's tortoise. The word *Turtle* was chosen rather than Tortoise as it's more widely used in the US.

CODING

Back to programming our turtles (and other sea creatures). You're going to create a dance where four dancers move towards each other and back to their starting positions.

⧉ **Remember** you can make the LEDs less bright (as in Chapter 8) by adding a set brightness block in the on start block.

THE DANCERS

First, you'll need to create **sprites** for the dancers.

> A **sprite** is a character in a video game that can
> move around on the screen. The micro:bit sprites
> are single pixels that can move around the display.

Create them one at a time so you know which is which. It's hard to tell a turtle from a lobster on the micro:bit screen!

The blocks for creating and working with sprites are under Advanced, in the Game section.

Introduce the turtle to your micro:bit.

Now add the seal.

The turtle and the seal both gain lobster partners.

☼ **Think** carefully about where each sprite is located on the micro:bit display.

The sprites of the partners need to turn to face the turtle and seal respectively. Sprites move in the direction they are facing. Sprites start off facing to the right, so you need to turn the partners 180° so they will move towards the turtle and seal.

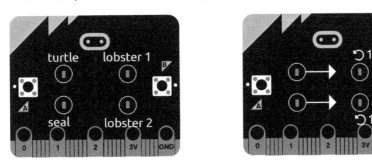

Add lobster1 opposite the turtle and rotate the sprite.

☼ **Think:** Would it make any difference if you turned left instead of right?

> Finally add `lobster2` opposite the `seal` and make him face his partner.

```
on start
    set turtle to      create sprite at x: 0  y: 1
    set seal to        create sprite at x: 0  y: 3
    set lobster1 to    create sprite at x: 4  y: 1
        lobster1  turn right by (°) 180
    set lobster2 to    create sprite at x: 4  y: 3
        lobster2  turn right by (°) 180
```

🚏 **Note:** You can't see the sprites rotating on the micro:bit display, but it will make a difference when you tell them to move!

MAKE THEM DANCE

Now let's make the creatures dance when you press button **A**. Code the next steps one at a time so you can clearly see how each dancer moves. It's tricky to distinguish between characters on the micro:bit so carefully watch the simulator as you regularly test your code.

🚏 **Remember** the turtle will move in the direction it is facing. It's already facing towards it's partner—lobster1.

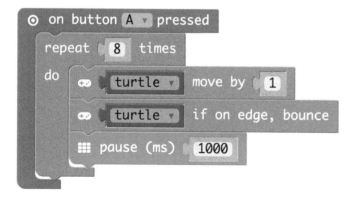

> First make the `turtle` move to its partner and back.

⏵ **Test** your code in the simulator and watch the turtle move.

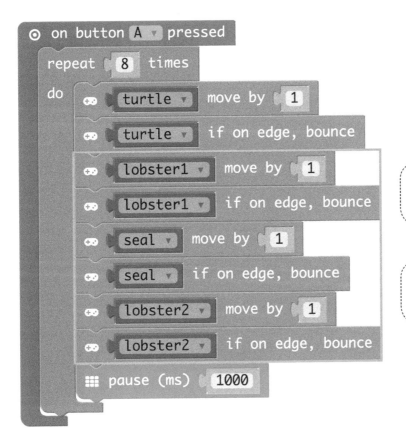

Now let lobster1 make the same dance move.

Time for the seal and lobster2 to join in too.

▶ **Test** your code in the simulator.

Did you notice how the lobsters moved towards to their partners and crossed over to the other side and then moved back to their starting positions?

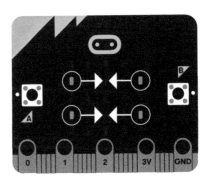

Click on the **snail** button in the simulator to slow things down if you need to **debug** your code.
(You may need to resize your window to see it!)

A **bug** is a mistake in code. Fixing bugs is called **debugging**.

A BIT OF HISTORY

WHY DO WE CALL ERRORS IN PROGRAMS *BUGS*?

Long before there was code it was common to call annoying engineering problems bugs which seems to come from the word bugbear (something that irritates you, originally a bad spirit). The inventor Thomas Edison frequently used the word bug when something wasn't working with his inventions (see, things don't always work the first time, even for famous inventors!)

"It has been just so in all my inventions. The first step is an intuition—and comes with a burst, then difficulties arise. This thing that gives out and then that—"Bugs" as such little faults and difficulties are called show themselves and months of anxious watching, study and labor are requisite before commercial success—or failure—is certainly reached."
Thomas Edison, 1878

Computer science pioneer Admiral Grace Hopper popularised the use of the term bug in computing. Her team even found a real bug, a moth, in an early computer!

SWAP PARTNERS

Now you're going to add code so when you press button **B** the dancers will swap with the dancer diagonally opposite this time. The dance partners will swap one pair at a time. The turtle and lobster2 will move in the dance like the diagram below.

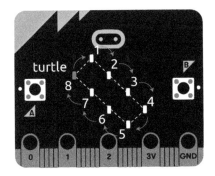 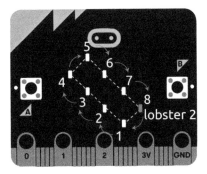

The turtle moves 4 spaces to swap with lobster2. It travels to lobster2's position. Then it repeats the steps to move back to its starting position. When the turtle bounces off the edge it changes direction at a 90° (degree) angle.

You're going to code each creature to make this dance move. First you need to make the creatures turn 45° (degrees) so they move diagonally.

They will turn outwards.

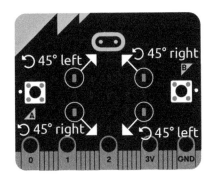

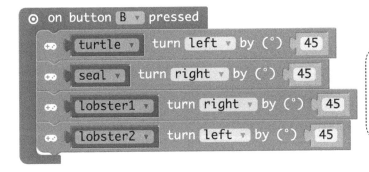

Add this code on button B pressed.
Take careful note which way each creature turns.

Now the creatures are all facing the right direction to start the dance.

...

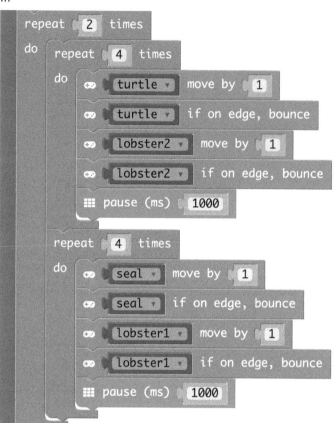

Add this code below the rotation blocks in the on button B pressed block.

This code will make the turtle swap with lobster2 and the seal swap with lobster1.

You need to repeat each creature's steps 4 times and repeat all the creatures' steps twice so they can return to their starting positions.

🔖 **Note:** The pause (ms) blocks are important otherwise it all happens too fast!

⋰ **Think:** Why does the turtle move 8 times?

Your creatures are back where they started but not facing the right direction. The following code will rotate them back to their starting positions so they are ready to repeat the dance when you press **B**.

...

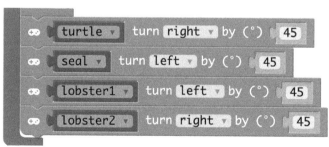

Add this code below the above blocks in the on button B pressed block.

Did you spot that they rotate in the opposite direction this time?

TRY IT OUT

▶ **Test** your dance on the simulator by clicking **A** for the first dance move and **B** for the second dance move.

⬇ **Download** your program and press **A** and **B** to make the creature dance on the micro:bit.

Find three other people (or 4 teddies, action figures, dolls, etc.) and use the micro:bit to teach them the dance moves.

🔖 **Tip:** Attach the battery pack so you can move around with the micro:bit. Secure the battery pack to the micro:bit using two loom bands. You could also use a MI:power board to power the micro:bit.

💬 *"'Thank you, it's a very interesting dance to watch," said Alice, feeling very glad that it was over at last: ...'*

MAKE IT YOURS

♣ Can you add more dance moves when other inputs are triggered? E.g. shake.

♣ How about adding music? Find some sheet music for a Quadrille on the website to get started.

♣ Can you change the code to slow the dance down so that you and your friends can follow it? Would you prefer to make it faster?

♣ When adding music the default tempo is 120 bpm (beats per minute). How you can change this to fit with the timing of the dance?

'"Herald, read the accusation!" said the King'

THE TRIAL

CHAPTER 12

FOLLOW THE STORY

📖 **Read** Chapter 11 *Who Stole the Tarts?* and Chapter 12 *Alice's Evidence* to find out what happens to Alice at the trial.

I n the trial, the Knave (Jack) is accused of stealing the tarts that have been baked by the Queen. The White Rabbit acts as the herald, announcing the witnesses and evidence using his trumpet and scroll.

💬 *The Queen of Hearts, She made some tarts, All on a summer day: The Knave of Hearts, he stole those tarts, And took them quite away!*

YOU WILL MAKE

In this chapter, you can make two separate projects. You will use the micro:bit to make the White Rabbit's trumpet and use it to read his scroll. If you have two micro:bits then you can make both projects at the same time.

For the scroll, you will use conductive ink, thread or tape to connect the micro:bit pins in different patterns. When you place the micro:bit on the scroll it will announce the next witness or piece of evidence in the trial.

For the trumpet, you will use a tone buzzer and program a fanfare on the micro:bit to announce the start of the trial.

YOU WILL LEARN

This chapter introduces **binary numbers** and working with multiple inputs. It provides an opportunity to work with **nuts and bolts** and the project can be completed using conductive ink, tape or thread. There's a chance to sew with **cross-stitch**, to make the scroll look amazing after practising on a **prototype** first.

YOU WILL NEED

ESSENTIALS

- ♣ micro:bit (battery pack and USB cable)
- ♣ Paper
- ♣ Loom bands
- ♣ Kitchen foil
- ♣ 4 crocodile clips
- ♣ 4 crocodile clip leads
- ♣ Scissors, sticky tape, glue, colouring pencils/felt tip pens

GET CREATIVE

- ♣ MI:power board (optional for the trumpet)
- ♣ 1 gold pencil, gold card (to make trumpet)
- ♣ 1 small buzzer, 2 M-to-F jumper wires, 2 more crocodile clips (to make trumpet)
- ♣ Pliers
- ♣ Conductive ink/tape/graphite pencil (to make paper scroll)
- ♣ Conductive thread, embroidery thread, cross-stitch fabric (to make fabric scroll)

TEMPLATES

🖶 **Print** a template on paper from the website for the scroll prototype, the final paper scroll and trumpet.

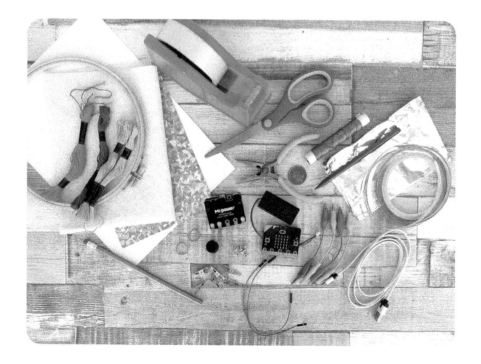

BINARY NUMBERS

You are going to use binary to tell the micro:bit which section of the White Rabbit's scroll it's reading. Computers use **binary** numbers to store data. Each place (digit/bit) in a binary number is either true (1) or false (0).

> **Binary** is a number system where the only digits are 0 and 1. Each place is worth twice as much as the previous place.

In our usual decimal counting system, each place is worth 10 times more than the previous place (1000 | 100 | 10 | 1). In binary, it's worth twice as much (8 | 4 | 2 | 1).

8s	4s	2s	1s
1	0	1	0
8 +	0 +	2 +	0

For example, 10 is made by adding (8x1) + (4x0) + (2x1) + (1x0). The binary number for 10 is 1010.

In this manner, with 3 places we can represent eight different numbers, including zero. The table on the right shows binary numbers (left column), how they are made (middle section) and their decimal number equivalent (right column).

Look through each row of the table and make sure that you understand how to get from binary numbers to decimal numbers.

Binary	4	2	1	Decimal
000	0	0	0	0
001	0	0	1	1
010	0	2	0	2
011	0	2	1	3
100	4	0	0	4
101	4	0	1	5
110	4	2	0	6
111	4	2	1	7

MAKING BINARY NUMBERS ON THE MICRO:BIT

Remember in the Drink Me, Eat Me project you made connections to the pins to create inputs. This time you're going to use combinations of inputs to create binary numbers. The micro:bit has three pins so let's say that Pin 2 is the units, Pin 1 is the 2s and Pin 0 is the 4s. When the pin is *off* (false) the digit is 0 and when the pin is *on* (true) the digit is 1.

Can you work out the binary number for the decimal numbers below? To get the binary number for three you need to add (4x0) + (2x1) + (1x1) which is 011.

Think about how binary numbers work and try to make other numbers in binary.

CODE THE BINARY NUMBERS

You'll use the binary numbers for 1-5 to trigger announcements on the White Rabbit's scroll.

001	1	The accusation: "The Knave of Hearts he ..."
010	2	The Mad Hatter
011	3	The Duchess's cook
100	4	Alice!
101	5	The letter

When you place the micro:bit on the scroll and press button **B** it will show an announcement with an image on the LED display. You'll work through this in steps. First, let's work out the numbers from the pins that are connected.

You're going to show the corresponding number on the micro:bit display so that you can make sure everything is working correctly. Later you'll code images to represent the numbers.

It's useful to build up projects step-by-step rather than jumping straight to the final solution to avoid having lots of code to debug at the same time.

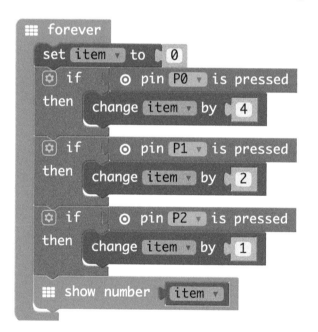

Add the following code to keep checking which pins are pressed and show the corresponding number.

Each time around the loop, `set item to 0` and then change it depending on which pins are pressed.

Note: If multiple pins are pressed then multiple numbers will get added to `item`.

Add a `show number` block to make sure everything is working as expected.

⤓ **Download** your program to the micro:bit.

It's difficult to test the pins in the simulator, so you're going to make a quick **prototype** to test your program.

> A **prototype** is a practice version of something. Prototyping is an important skill in developing your designs.

TESTING

Use kitchen foil to create conductive touch pads on the prototype template.

You will need 4 crocodile clip leads to connect the pins to the prototype as shown below.

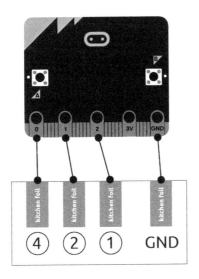

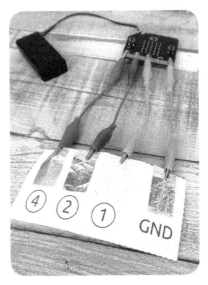

Now you can make the binary numbers appear on the micro:bit display by connecting different combinations of pins. Touch ground (GND) and Pin 2 to make the number 1. Use two hands to make the connections. Try make all the numbers from 1 - 5.

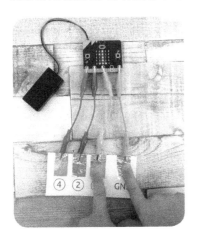

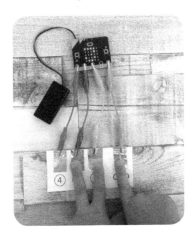

1	Just Pin 2
2	Just Pin 1
3	Pin 2 and 1
4	Just Pin 4
5	Pin 4 and 1

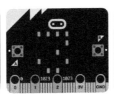

⁙ **Challenge:** Which pins do you need to connect to make 6 and 7?

When you're happy that your prototype is working you can make the scroll. Test the scroll with the numbers before coding the trial pictures to help debug any connections.

PREPARE THE MICRO:BIT

First you need to prepare the micro:bit to make reading the scroll a little easier.

Fit a crocodile clip to Pin 0, 1, 2 and GND at right angles to the micro:bit. Attach the battery pack to the micro:bit using two loom bands.

Make sure the crocodile clips are nice and straight and don't touch each other.

MAKE A PAPER SCROLL

Use the template for the paper scroll to create conductive patterns that make the same connections that you made with your prototype.

Tip: There's enough space on your scroll for items 6 and 7 if you want to decide what happens at the end of the trial (we never find out in the book!)

Use either conductive ink, conductive tape, a graphite pencil or kitchen foil to make the patterns conductive so that the micro:bit pins can be connected.

Decorate your scroll.

Now test your binary numbers on your scroll by holding the micro:bit on the patterns you made. Make sure the pins line up and a good connection is made with GND every time.

Troubleshooting: If it's not working, make sure the correct connections are touching and making good contact.

CODE THE ACCUSATIONS

The White Rabbit first unrolls the parchment scroll and reads out the accusation. It's time to add an image on the micro:bit display instead of showing the number.

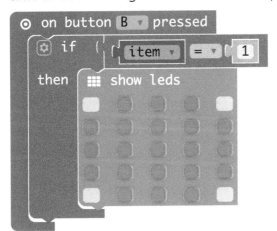

Add a logic block to create a *jam tart* on show leds if the variable item is equal to 1.

The White Rabbit calls the first witness, the Mad Hatter. Remember to click the **gear** icon to add more else if blocks.

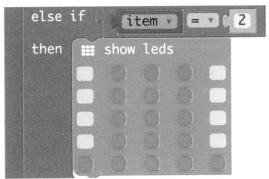

This is triggered by connecting Pin 1 only to GND on the micro:bit to create the number 2.

Next is the Cook.

Add the *sneeze* to represent The Cook on number 3.

Now it's Alice's turn to face the court.

...

Show a string *Alice!* on number 4.

Finally comes a letter written by the prisoner (actually it's a set of verses).

...

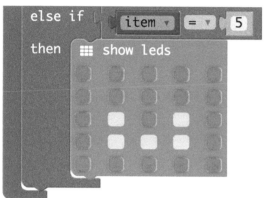

Design a letter on the show leds block to represent the Prisoner on number 5.

⤓ **Download** your program and test it on your scroll.

CRAFT

MAKE A FABRIC SCROLL

You can also make the scroll using cross-stitch fabric and sew the patterns with conductive thread. Place the fabric over the template so that you can see the patterns through the fabric and trace them using a pencil onto the fabric.

🚏 **Tip:** It's best to use one continuous piece of conductive thread for each pattern. Test after sewing the first pattern to make sure everything works.

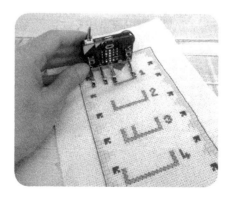

MAKE IT YOURS

Share your make #techalice

♣ Use non-conductive embroidery thread to decorate the scroll to look like a Victorian sampler.

♣ Can you improve the images that are shown on the micro:bit for each announcement?

♣ You haven't used the binary numbers 6 or 7. Can you work out how to represent them in binary and add two more items to the scroll? The book ends prior to the final verdict so we don't find out what happens. What do you think?

MAKE TWO: THE TRUMPET

💬 *"'Herald, read the accusation!' said the King. On this the White Rabbit blew three blasts on the trumpet, ..."*

Before making announcements to the court the White Rabbit plays three notes on his trumpet.

CODING

Code the micro:bit to display a heart as shown on the banner of the White Rabbit's trumpet. The micro:bit will be upside down when it's attached to the trumpet.

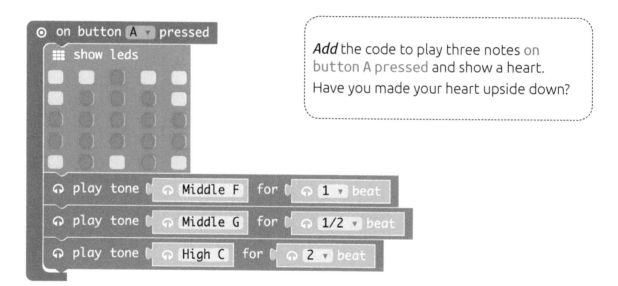

Add the code to play three notes on button A pressed and show a heart.
Have you made your heart upside down?

▶ **Test** your code in the simulator. You can change the notes if you wish.

⬇ **Download** the program to your micro:bit and attach the battery pack using two loom bands. Alternatively use the MI:power board to power the micro:bit and play the sound.

Make a trumpet using an unsharpened pencil and card (gold would make it look more authentic). You'll also need a small buzzer, 2 jumper wires, 2 crocodile clips and some sticky tape.

Use the template to trace and cut out a bell for the trumpet on gold card.

Push the pins of the buzzer into one end of both the jumper wires. Slot the unsharpened end of the pencil in-between the wires and secure with sticky tape as shown below.

Use pliers to crimp a crocodile clip to each of the pins of the jumper wires to hold them in place.

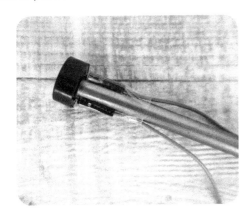
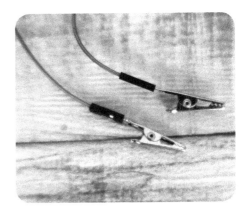

Wrap the bell around the pencil to conceal the buzzer. Secure with sticky tape. Coil the jumper wires around the pencil so you are left with just the crocodile clips dangling down. Use sticky tape to stop the wires from uncoiling.

CONNECTING

Attach one crocodile clip to Pin 0 and the other to GND. Press button **A** to play the trumpet and begin the trial. The heart should display right side up!

MAKE IT YOURS

Share your make #techalice

♣ Can you program a more interesting tune for the White Rabbit's trumpet?

♣ Do you want the heart to display on the trumpet all of the time or change to a music note icon when the music is playing?

💬 *'"Who cares for you?" said Alice, (she had grown to her full size by this time.) "You're nothing but a pack of cards!" At this the whole pack rose up into the air, and came flying down upon her: she gave a little scream, half of fright and half of anger, and tried to beat them off, and found herself lying on the bank, with her head in the lap of her sister, who was gently brushing away some dead leaves that had fluttered down from the trees upon her face.'*

'"Oh, I've had such a curious dream!" said Alice, and she told her sister, as well as she could remember them, all these strange Adventures of hers ...'

ADDITIONAL PROJECTS

MORE PROJECTS FOR YOUR BBC MICRO:BIT

Finished all of the projects? Below are some extra challenges that allow you to make use of the skills you have developed. Some of these projects allow you to extend projects in the first few chapters with skills that you have learnt later in the book.

- ♣ Create animated watch hands for your pocket watch.
- ♣ Use `if..else` blocks instead of just `if` blocks in the Drink Me, Eat Me project.
- ♣ Use play dough or real mushroom halves to create inputs to grow and shrink Alice by one pixel at a time using a variable to keep track of her height.
- ♣ Add a sneeze sound effect to the Pepper Box project. Which musical notes sound most as a sneeze?
- ♣ Add sound to projects that didn't have sound, such as the Lobster Quadrille or the Cheshire Cat animation.
- ♣ Sew more e-textiles accessories to make inputs for projects. Sew the cake, bottle and mushroom with conductive thread and press-studs.
- ♣ Improve on your animations or add animations to projects that didn't have any.
- ♣ Make a wearable for the Lobster Quadrille with a LED that flashes in time with the beat.
- ♣ Use binary to light up LEDs to count to 7 to show the Queen's anger levels.
- ♣ At the very end of *Alice's Adventures in Wonderland,* the cards all fly up into the air. Can you use the on free fall event block to detect when the micro:bit is falling and scroll the text *The End*. Be careful not to drop the micro:bit.
- ♣ Use the story to inspire your own project to make a new game, gadget, wearable, animation, prop or decoration.

⊕ **Find** three additional bonus projects for *micro:bit in Wonderland* on the website (alice.techagekids.com).

Parents and teachers don't forget to share your thoughts and the children's creations on social media using the hashtag *#techalice*.

Find us on: 📘 🐦 📷 📌 G+ ▶️

Share your make #techalice

📑 **Note:** If you're not old enough to post online, ask an adult for help. Remember to keep your personal information safe.

A PERSONAL NOTE FROM THE AUTHORS

We hope you've enjoyed our trip down the rabbit hole. We've certainly enjoyed sharing our love of craft and tech with you in Wonderland.

We encourage you to keep making things that are fun, useful, entertaining, beautiful and curious. Use the skills you've learnt in this book to make things that appeal to your style and interests. Remember that it's okay if projects don't turn out as planned. You often learn the most when things go wrong or take an unexpected turn. Enjoy the adventure.

Keep the wonder and maybe we'll see you through the looking-glass one day soon.

Happy digital making,

Elbrie and Tracy

NEXT BOOK IN THE SERIES

MICRO:BIT THROUGH THE LOOKING-GLASS

Continue your adventure with the BBC micro:bit in Wonderland. *"... It's a plan of my own invention. You may try it if you like."* said the Knight. Join Alice as she enters Wonderland again through the looking-glass in the drawing room.

The second coding and craft project book, *micro:bit through the Looking-Glass*, builds on skills learnt from Book 1 in the *micro:bit in Wonderland* series.

The book introduces slightly more complex coding concepts and electronics but is still suitable for beginners aged 9 and over.

The projects provide an opportunity to work with a servo motor and programmable multicoloured LED lights and further develop craft and making skills using construction toys, e-textiles and papercraft.

The book offers an opportunity to learn more about wearables, electronic games, e-textiles, electronics circuits, animation, science, usability and much more.

Look out for *micro:bit through the Looking-Glass* on alice.techagekids.com in 2018.

Lightning Source UK Ltd.
Milton Keynes UK
UKHW05f1839200418
321391UK00002B/20/P